Lecture Notes in Computer Science 15454

Founding Editors

Gerhard Goos
Juris Hartmanis

Editorial Board Members

Elisa Bertino, *Purdue University, West Lafayette, IN, USA*
Wen Gao, *Peking University, Beijing, China*
Bernhard Steffen, *TU Dortmund University, Dortmund, Germany*
Moti Yung, *Columbia University, New York, NY, USA*

The series Lecture Notes in Computer Science (LNCS), including its subseries Lecture Notes in Artificial Intelligence (LNAI) and Lecture Notes in Bioinformatics (LNBI), has established itself as a medium for the publication of new developments in computer science and information technology research, teaching, and education.

LNCS enjoys close cooperation with the computer science R & D community, the series counts many renowned academics among its volume editors and paper authors, and collaborates with prestigious societies. Its mission is to serve this international community by providing an invaluable service, mainly focused on the publication of conference and workshop proceedings and postproceedings. LNCS commenced publication in 1973.

Josu Doncel · Anne Remke · Daniele Di Pompeo
Editors

Computer Performance Engineering

20th European Workshop, EPEW 2024
Venice, Italy, June 14, 2024
Revised Selected Papers

Editors
Josu Doncel
University of the Basque Country
Leioa, Spain

Anne Remke
University of Münster
Münster, Germany

Daniele Di Pompeo
University of L'Aquila
L'Aquila, Italy

ISSN 0302-9743 ISSN 1611-3349 (electronic)
Lecture Notes in Computer Science
ISBN 978-3-031-80931-6 ISBN 978-3-031-80932-3 (eBook)
https://doi.org/10.1007/978-3-031-80932-3

© The Editor(s) (if applicable) and The Author(s), under exclusive license
to Springer Nature Switzerland AG 2025

This work is subject to copyright. All rights are solely and exclusively licensed by the Publisher, whether the whole or part of the material is concerned, specifically the rights of translation, reprinting, reuse of illustrations, recitation, broadcasting, reproduction on microfilms or in any other physical way, and transmission or information storage and retrieval, electronic adaptation, computer software, or by similar or dissimilar methodology now known or hereafter developed.
The use of general descriptive names, registered names, trademarks, service marks, etc. in this publication does not imply, even in the absence of a specific statement, that such names are exempt from the relevant protective laws and regulations and therefore free for general use.
The publisher, the authors and the editors are safe to assume that the advice and information in this book are believed to be true and accurate at the date of publication. Neither the publisher nor the authors or the editors give a warranty, expressed or implied, with respect to the material contained herein or for any errors or omissions that may have been made. The publisher remains neutral with regard to jurisdictional claims in published maps and institutional affiliations.

This Springer imprint is published by the registered company Springer Nature Switzerland AG
The registered company address is: Gewerbestrasse 11, 6330 Cham, Switzerland

If disposing of this product, please recycle the paper.

Preface

The European Performance Engineering Workshop (EPEW) aims to bring together researchers working on performance evaluation of systems from theoretical and practical viewpoints. The concept of performance in EPEW 2024 was considered in its broadest sense including the notions of Quality of Service, scalability, reliability, availability and systems management, among others.

The 20th edition of EPEW was held on the 14th of June 2024 in Venice, and it was co-located with the conference ACM SIGMETRICS/IFIP PERFORMANCE 2024. This edition was a very fruitful workshop with numerous presentations dealing with performance evaluation research problems.

We acknowledge the organizers of the conference ACM SIGMETRICS/IFIP PERFORMANCE 2024 (the General Chairs Andrea Marin and Michele Garetto as well as the Workshop Chairs Dieters Fiems and Valeria Cardellini and the Local Arrangement Chairs Diletta Olliaro and Sabina Rossi) for their help in the organization of this edition of EPEW. We also thank Mirco Tribastone for accepting the invitation to give a keynote talk in this edition of EPEW, which was entitled 'Software Performance Modeling for the Cloud: An Overview'. Furthermore, we acknowledge the work of the Program Committee, which was formed by 28 top-level researchers from different countries.

In this edition, there was a two-phase review process. First, we solicited short papers of at most 7 pages for presentation at the conference; these short papers will not be published. Then, authors of accepted short papers were invited to submit a full paper with at most 15 pages. Accepted full papers appear in this volume. Initially, we received 15 short papers, from which 13 were accepted for presentation at the conference. Besides, we received 12 full paper submissions and 10 of them were accepted for publication in this volume. Each paper was single-blindly reviewed in each of the rounds by at least three reviewers.

Finally, we would like to thank all the authors that contributed to this workshop, for either submitting a paper to the workshop or giving a talk during the event, or submitting a full paper for publication in this volume.

October 2024

Daniele Di Pompeo
Josu Doncel
Anne Remke

Organization

General Chair

Josu Doncel — University of the Basque Country, Spain

Program Committee Chairs

Daniele Di Pompeo — University of L'Aquila, Italy
Anne Remke — University of Münster, Germany

Program Committee

Salvador Alcaraz — Miguel Hernández University, Spain
Elvio Gilberto — Amparore University of Turin, Italy
Paolo Ballarini — CentraleSupélec, France
Marco Bernardo — University of Urbino, Italy
Laura Carnevali — University of Florence, Italy
Davide Cerotti — Università del Piemonte Orientale, Italy
Carina Da Silva — University of Münster, Germany
Dieter Fiems — Ghent University, Belgium
Matthew Forshaw — Newcastle University, UK
Jean-Michel, Fourneau — Université de Versailles St Quentin, France
Pedro Pablo Garrido Abenza — Miguel Hernández University, Spain
Antonio Gomez — Complutense University of Madrid, Spain
Marco Gribaudo — Politecnico di Milano, Italy
Nikolas Herbst — University of Würzburg, Germany
András Horváth — University of Turin, Italy
Alain Jean-Marie — Inria, France
Carlos Juiz — University of the Balearic Islands, Spain
Lasse Leskelä — Aalto University, Finland
Francesco Longo — University of Messina, Italy
Marco Paolieri — University of Southern California, USA
Nihal Pekergin — Univ. Paris-Est Creteil, France
Tuan Phung-Duc — University of Tsukuba, Japan
Agapios Platis — University of the Aegean, Greece
Marco Scarpa — University of Messina, Italy

Markus Siegle   University of Munich, Germamy
Nigel Thomas   University of Newcastle, UK
Catia Trubiani   Gran Sasso Science Institute, Italy
Joris Walraevens   Ghent University, Belgium

Additional Reviewers

Lukas Beierlieb

Contents

Design and Analysis of Distributed Message Ordering over a Unidirectional Logical Ring 1
    Ye Liu, Paul Ezhilchelvan, and Isi Mitrani

The Omnibus Java Library: Efficient Synthesis of Optimal Signal Schedules for Multimodal Intersections 14
    Nicola Bertocci, Laura Carnevali, Leonardo Scommegna, and Enrico Vicario

Performance Evaluation of Beaconing Schemes for Vehicular Platooning 29
    Hassan Laghbi and Nigel Thomas

Implementations Based Evaluation of No-Wait Approach for Resolving Conflicts in Databases 45
    Yingming Wang, Paul Ezhilchelvan, Jack Waudby, and Jim Webber

Performance Evaluation of Smart Bin Systems Using Markovian Agents for Efficient Garbage Collection 60
    Enrico Barbierato, Alice Gatti, Marco Gribaudo, and Mauro Iacono

Approximation of First Passage Time Distributions of Compositions of Independent Markov Chains 75
    András Horváth, Marco Paolieri, and Enrico Vicario

Under the Space Threat: Quantitative Analysis of Cosmos Blockchain 91
    Daria Smuseva, Ivan Malakhov, Andrea Marin, Carla Piazza, and Sabina Rossi

A Lumped CTMC for Modular Rewritable PN 106
    Lorentzo Capra and Marco Gribaudo

Analytical Modelling of Asymmetric Multi-core Servers 121
    M. Gribaudo and T. Phung-Duc

Robust Streaming Benchmark Design in the Presence of Backpressure 137
    Iain Dixon, Matthew Forshaw, and Joe Matthews

Author Index 153

Design and Analysis of Distributed Message Ordering over a Unidirectional Logical Ring

Ye Liu, Paul Ezhilchelvan[(✉)], and Isi Mitrani

School of Computing, Newcastle University, NE4 5TG Newcastle, UK
{Y.Liu197,paul.ezhilchelvan,isi.mitrani}@newcastle.ac.uk

Abstract. Several servers generate and disseminate messages which must be processed in the same order by all of them. A ring protocol is proposed, where a folder carrying messages circulates in one directiontionting queueing model is analysed in the steady state and an approximate solution is developed, allowing the computation of performance measures. This is applied to some example systems and the results are compared with simulations.

1 Introduction

Message ordering involves distributed servers processing messages in an identical order. This is long known as a fundamental requirement for building crash tolerant services through replicated processing [11]. The messages being ordered typically contain requests, e.g. file updates, to a replicated file service. A client sends its request to any one of the replicated servers (typically on 3 or 5 distinct hosts), which then disseminates the request to all other servers so that updates everywhere are performed in a mutually consistent manner. Apache Zookeeper [6] is an industrial strength replicated service used in many practical applications.

There is an extensive literature on ordering protocols (see [2] for a survey). Most take a centralised approach, including Zookeeper and its newer version RAFT [10], in order to minimise ordering latency. Here, one replica is designated as the leader to whom all others send their messages for ordering; leader then does two multicasts for disseminating and then confirming its ordering choice. Most networked systems do not readily support group or IP multicast whereby a sender can transmit a message to multiple destinations in one operative step, but rather require that the sender carry out multiple unicasts, one for each destination. This aspect of multiple unicasting increases network traffic and the load placed on the leader [3].

More recently, interest has focused on high throughput and scalable data processing (even at the expense of latency). The major constraint in achieving this appears to be the network capacity [13]. Consequently, alternatives to multiple unicasting have been explored and the prominent approach is to arrange server replicas as a chain. Each server sends messages only to its neighbour in one direction, and receives them only from its neighbour in the other direction [4]. The resulting structure is a leader-free, decentralised ring. It is argued in [5] that the message ordering over a ring structure can offer scalable throughput that closely matches the network throughput.

In this paper, we propose a new ring protocol wherein a server disseminates its messages in batches. A folder circulates continuously around the ring, visiting each server in turn. It contains a fixed number of message slots dedicated to each server. Messages are numbered sequentially. When the folder arrives at a server, all messages in it are copied and ordered according to their sequence numbers; the server then removes its own messages from the folder and fills its slots with new ones waiting to be disseminated. Each of the newly loaded messages is sequentially numbered by working out the largest sequence number in the received folder. The folder is then sent to the next server in the ring.

If servers on a ring are allowed to transmit messages at any time (i.e., without a folder), some form of fairness control is essential. Without such control, a heavily loaded server may choose to give priority to its own messages, at the expense of those sent to it for forwarding. This issue was recognised and addressed in [5] by means of a rather complex algorithm. Our protocol imposes fairness in an easily implementable manner by means of the circulating folder. The price paid for that simplicity is that servers are constrained to transmit only when visited by the folder, and each transmission is limited by the number of slots allocated to them.

We analyse the queueing behaviour of the folder protocol and provide an approximate solution that allows us to compute performance measures. Such an analysis has not, to our knowledge, been done before. There is a resemblance between our circulating folder and a 'polling server' which visits a number of stations. While there is a large volume of work on polling systems (e.g., see [12] for a good survey), none of the existing results apply to our model. The main reason for this is that the time the folder remains at a server depends on the total number of occupied slots, i.e. not only on the state of the current server's queue, but also on the queues of all other servers.

The single folder version can be easily extended to multiple circulating folders. However, the analysis would then become considerably more complicated and is therefore left for future work. Another issue that we do not consider here relates to the possibility of server crashes. After a breakdown, the ring structure would need to be reconfigured and a new folder must be initialised. During this recovery process, servers would communicate with each other in the normal manner and can use any of the algorithms proposed in [5,7,9,10]. Recovery would be possible, as long as no more than one server crashes between successive reconfigurations. For our study, we assume that all servers are reliable.

Another, more distantly related work is [8], where a combination of a leader and a ring is proposed. There is also a class of protocols where a logical ring is used for rotating the leadership role among servers [2].

The model is described in Sect. 2. Section 3 develops the approximate solution, while Sect. 4 presents several examples.

2 The Model

We consider a system with N service stations, numbered 1, 2, ..., N. They communicate with each other by means of a folder which travels in one direction only: station 1 sends it to station 2, station 2 sends it to station 3, ..., station N sends it to station 1. The folder contains spaces referred to as 'slots', each of which may carry one message. Station i has k_i slots within the folder reserved exclusively for its use ($i = 1, 2, \ldots, N$). Thus the maximum number of messages carried by the folder is $K = k_1 + k_2 + \ldots + k_N$.

Messages requiring transmission arrive at station i according to an independent Poisson process with rate λ_i, and join a separate FIFO queue, Q_i. The dispatching protocol works as follows. When the folder comes to station i, up to k_i messages from Q_i are loaded into it, in their order of arrival; if there are fewer than k_i messages in Q_i at that moment, then some of the reserved slots would be unused. As the folder visits the other stations in turn, each of them copies Q_i's messages and loads up to its reserved number of slots with messages from its own queue. When the folder returns to Q_i, the messages loaded on the previous visit are considered to have been delivered and are deemed to have departed from the system. A new batch of messages from Q_i is loaded. The same protocol applies to all stations.

Let a be the average time it takes to copy or load one slot. If all slots are being used, the folder is delayed for an average interval of aK at each station. The average time, t_i, to transfer the folder from station i to the next station may also depend on the number of occupied slots. If all K allocated slots are occupied, that average has the form $t_i = bK + \beta_i$, where b and β_i are given constants. Thus, the maximum average cycle time, $\bar{T}$, i.e. the average interval between two consecutive visits of the folder to station i, is given by

$$\bar{T} = NK\alpha + \beta ,\qquad(1)$$

where $\alpha = a + b$ and $\beta = \beta_1 + \beta_2 + \ldots + \beta_N$. During such a cycle, k_i messages originating at station i are delivered and leave the system. Of course, if there are unused slots, then both the cycle time and the number of departures would be lower.

It is intuitively clear that the system as a whole is stable if, at all stations, the average number of messages that arrive during a maximum cycle is lower than those that can be delivered:

$$\lambda_i \bar{T} < k_i \ ; \ \ i = 1, 2, \ldots, N .\qquad(2)$$

If, when the folder arrives at station i, there are fewer than k_i messages in Q_i, some of the slots assigned to that station will be unused. Moreover, some of the slots allocated to other stations may be travelling empty. Thus, the time the folder remains at station i on each visit depends not only on the current state of Q_i, but also on the previous states of the other queues. This is the crucial difference between the circulating folder in this system, and a polling server.

The complex interdependencies between the queueing processes at different stations mean that an exact analysis of this model is intractable. We therefore propose an approximate solution that will enable us to compute reasonable estimates of performance measures.

3 Fixed-Point Approximation

Consider station i in isolation, with Q_i evolving in a stationary environment defined by the other stations. That is, Q_i is treated as an M/M/1 queue with Poisson arrivals and state-dependent 'bulk services'. There is a sequence of 'departure moments', when several messages leave the queue simultaneously. These moments correspond to the visits of the folder at station i. The rates at which they occur, and the number of departures that take place, depend on the state of the queue and are modulated by the environment.

In order to simplify the notation, we shall omit the index i from the arrival rate, slot allocation and steady-state probabilities of Q_i. The behaviour of that queue is described as follows. If there are at least k messages present, then the next departure moment occurs at rate μ_k and the number of messages departing at that moment is k. If there are j messages present ($j = 1, 2, \ldots, k-1$), then the next departure moment occurs at rate μ_j and the number of departures is j; the queue would then be emptied.

The environment is defined by the average numbers of slots in the folder that are occupied by the other stations. Those averages will be referred to as 'occupancies' and will be denoted by s_m, $m \in 1, 2, \ldots, N$, $m \neq i$. The sum, S, of the other stations' occupancies is assumed fixed and is the environment in which Q_i evolves.

For a given value of S, if j messages depart from Q_i, the average cycle length of the folder would be

$$T_j = N(j+S)\alpha + \beta \ ; \ \ j = 1, 2, \ldots, k \ , \tag{3}$$

where α and β are the parameters that appear in (1). Hence, the rates at which departure moments occur at Q_i can be expressed as

$$\mu_j = [N(j+S)\alpha + \beta]^{-1} \ ; \ \ j = 1, 2, \ldots, k \ . \tag{4}$$

These rates are state-dependent when there are fewer than k messages in Q_i, and become state-independent when there are k or more messages present.

The proposed approximation consists in assuming that Q_i in isolaton behaves as a Markov process, with arrival rate λ and bulk departure rates given by (4). That is, for a given environment, the intervals between consecutive visits of the folder to Q_i are assumed to be distributed exponentially with state-dependent parameters μ_j.

Let π_j be the steady-state probability that there are j messages in queue Q_i. These probabilities satisfy the following set of balance equations.

$$\lambda \pi_0 = \sum_{j=1}^{k} \mu_j \pi_j , \qquad (5)$$

$$(\lambda + \mu_j)\pi_j = \lambda \pi_{j-1} + \mu_k \pi_{j+k} \ ; \ j = 1, 2, \ldots, k-1 , \qquad (6)$$

and for all states where $j \geq k$,

$$(\lambda + \mu_k)\pi_j = \lambda \pi_{j-1} + \mu_k \pi_{j+k} \ ; \ j = k, k+1, \ldots . \qquad (7)$$

It is known that, if the departure rates are state-independent, the steady-state distribution of a queue with bulk services is geometric (e.g., see [1]). Here we cannot expect a geometric distribution because the rates are state-dependent. However, the distribution of Q_i turns out to have a geometric tail. Specifically, we can find a solution to equations (7) (for $j > k$), of the form

$$\pi_j = C z_0^j \ ; \ j = k, k+1, \ldots \qquad (8)$$

where C and z_0 are some positive constants. Indeed, substituting (8) into (7), we find that the equations are satisfied as long as z_0 is a zero of the polynomial of degree $k+1$

$$P(z) = \lambda(1-z) - \mu_k z(1-z^k) . \qquad (9)$$

In addition, in order that we may obtain a probability distribution, z_0 must satisfy $|z_0| < 1$.

Note that $P(0) > 0$ and $P(1) = 0$. Hence, $P(z)$ has a real zero, z_0, in the interval $(0, 1-\epsilon)$. Moreover, it can be shown that $P(z)$ has no other zeros in the interior of the unit disk. That follows from Rouche's theorem.

The probabilities π_j, for $j = 0, 1, \ldots, k-1$, and the constant C, are determined from equations (6) and (5), and from the requirement that the sum of all probabilities must be 1. The easiest way to perform that computation is to start by setting $C = 1$ and $\pi_k = 1$, so that $\pi_{k+j} = z_0^j$. Use equations (6) to determine $\pi_{k-1}, \pi_{k-2}, \ldots, \pi_1$ in turn, and equation (5) to determine π_0. Finally, normalize the computed 'probabilities', multiplying each of them by the constant

$$C = \left[\sum_{j=0}^{k-1} \pi_j + \frac{\pi_k}{1-z_0} \right]^{-1} . \qquad (10)$$

Having determined the distribution of Q_i, the corresponding occupancy, s, (i.e. the average number of messages it occupies in the folder), is obtained from

$$s = \sum_{j=1}^{k-1} j \pi_j + \frac{k \pi_k}{1-z_0} . \qquad (11)$$

That occupancy has an upper bound k, and a lower bound which would be achieved if all other queues were always empty.

The average number of messages, L, present in Q_i, is given by

$$L = s + \frac{\pi_k z_0}{(1-z_0)^2} \, . \qquad (12)$$

Now we need to take into account the interdependency between the N stations. The above computational procedure yields the occupancy of one station, given the occupancies of the other stations. That occupancy in turn influences the other stations by taking part in *their* environments.

Suppose that we start by assuming some occupancies, $(s_2, s_3, \ldots, s_N)$, for stations 2, 3, ..., N. Applying (11) we determine the occupancy of station 1, s_1. Then, using the new s_1, together with the old $s_3, \ldots, s_N$, we can compute a new value, s_2, for the occupancy of station 2. The new s_1 and s_2, together with the old $s_4, \ldots, s_N$, determine a new s_3, and so on. After N such steps, the values of all occupancies are renewed.

We can summarize the above N-step procedure by denoting it as a function, f, which transforms an old vector of occupancies, $\mathbf{s}^{old} = (s_1^{old}, s_2^{old}, \ldots, s_N^{old})$ into a new vector of occupancies, $\mathbf{s}^{new} = (s_1^{new}, s_2^{new}, \ldots, s_N^{new})$:

$$\mathbf{s}^{new} = f(\mathbf{s}^{old}) \, . \qquad (13)$$

The model solution is provided by the fixed point, $\mathbf{s}^*$, of the function f. That is a 'mutually consistent' vector of occupancies which does not change when used to define the environments of different stations:

$$\mathbf{s}^* = f(\mathbf{s}^*) \, . \qquad (14)$$

Intuitively, such a fixed point exists because f is continuous and the set of occupancy vectors is bounded and convex. Moreover, while not having a proof, we conjecture that the fixed point is unique. A referee has suggested a possible approach to proving that conjecture, but we feel that such a proof is outside the scope of the present study.

The point $\mathbf{s}^*$ can be computed by iterating the transformation f. Start with an initial guess, $\mathbf{s}_0$, e.g. $\mathbf{s}_0 = (k_1, k_2, \ldots, k_N)$ (this is the initial guess that we have usually adopted). At the nth iteration, compute

$$\mathbf{s}_n = f(\mathbf{s}_{n-1}) \; ; \; n = 1, 2, \ldots \, . \qquad (15)$$

Stop when two consecutive iterations are sufficiently close to each other and return the resulting performance measures $(L_1, L_2, \ldots, L_N)$.

3.1 Special Cases

The complexity of the solution is reduced considerably when the system is symmetric. If the arrival rates and slot allocations are the same at all stations, $\lambda_i = \lambda$ and $k_i = k$ for all i, then the occupancies are also the same, $s_i = s$, and the environment of any isolated station is defined by a total occupancy, $(N-1)s$, of the other stations. For a given value of s, the state-dependent departure rates (4) become

$$\mu_j = [N(j + (N-1)s)\alpha + \beta]^{-1} \; ; \; j = 1, 2, \ldots, k \; . \tag{16}$$

The mutually consistent occupancy, s, is produced by the fixed-point iterations (15), which are now in terms of a single variable.

If, in a symmetric system, all stations are allocated one slot each, the solution can be obtained in closed form. In this case messages depart singly and the occupancy variable for every queue, s, is the probability that the queue is not empty. The state-independent departure rate is given by (16), with $j=1$:

$$\mu_1 = [N(1 + (N-1)s)\alpha + \beta]^{-1} \; . \tag{17}$$

Since the probability that the queue is not empty is equal to $s = \lambda/\mu_1$, we can write the fixed-point equation for s as

$$s = \lambda[N(1 + (N-1)s)\alpha + \beta] \; . \tag{18}$$

When the stability condition (2) is satisfied, i.e. when $\lambda(N^2\alpha + \beta) < 1$, this equation has a unique solution in the interval (0,1):

$$s = \frac{\lambda(N\alpha + \beta)}{1 - \lambda N(N-1)\alpha} \; . \tag{19}$$

That solution provides an estimate for the average queue size at each station:

$$L = \frac{s}{1-s} \; . \tag{20}$$

The solution of an asymmetrical system depends on how many stations have different parameters. Suppose, for example, that stations 2, 3, ..., N are identical, with arrival rates equal to λ and slot allocations k each, while station 1 is different, with arrival rate λ_1 and slot allocation k_1. Then there would be one environment for station 1, defined by the total occupancy, $(N-1)s$, of stations 2, 3, ..., N, and another environment for any of the other stations, with total occupancy $s_1 + (N-2)s$ that includes station 1. The fixed-point iterations would compute a two-element vector (s_1, s).

4 Examples

We have experimented first with a symmetric system containing 5 stations, all having the same parameters. The processing and communication parameters are fixed at $a + b = 10^{-3}$ secs and $\beta = 10^{-5}$ secs, respectively. The arrival rate is varied, and the average number of messages present at one station, L, is estimated by the model and by simulation, for purposes of comparison.

In the first example, just one slot is allocated to each station: $k_i = 1$ for all i. The estimated values of L are computed according to the closed-form expressions (19) and (20).

The maximum cycle time is about 0.025 secs, which means that, for stability, the arrival rate at each station must be less than 40 messages per second. In the simulation, it was assumed that the intervals between a visit of the folder to one station and the visit to the next station are distributed exponentially. When a total of S slots are occupied, the average length of that interval is $v = S\alpha + \beta/N$. Then the simulation generates an exponentially distributed random number with mean v. When all slots are occupied, the mean is 0.005 secs. Hence, the simulated maximum cycle has a 5-phase Erlang distribution with mean 0.025 secs, whereas in the model it is distributed exponentially with the same mean. This discrepancy adds to the approximate nature of the results.

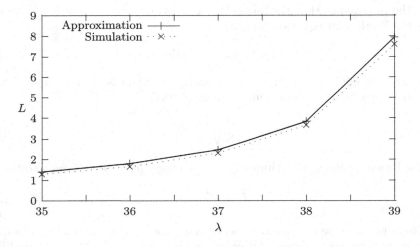

Fig. 1. Symmetric system. Increasing arrival rate, $k = 1$

In Fig. 1, the arrival rate per station increases from $\lambda = 35$ to $\lambda = 39$. As expected, L increases non-linearly. It would have a vertical asymptote close to $\lambda = 40$. The approximation is consistently pessimistic, possibly because the exponential distribution has a higher coefficient of variation than the Erlang. However, the estimated average queue sizes are very accurate. Their relative errors, compared with the simulated values, are on the order of 5%.

For each value of λ, ten independent simulation runs were made for the purpose of computing the 95% confidence intervals. During each run, a total of 100000 messages arrived into the system. The half-widths of the resulting confidence intervals are displayed in Table 1. We observe that all but the first two approximated values of L are within the corresponding confidence intervals.

Table 1. Confidence intervals for estimates in Fig. 1

λ	35	36	37	38	39
L approximated	1.41	1.81	2.48	3.84	7.94
L simulated	1.30	1.68	2.43	3.66	7.70
conf. interval half-width	0.03	0.06	0.20	0.46	1.20

It is interesting to consider the effect of increasing the slot allocation per station, k. In a symmetric system, the stability condition (2) can be written as

$$\lambda < \frac{k}{N^2 k \alpha + \beta} \,. \tag{21}$$

Since the value of β is typically close to 0, that condition is almost independent of k and is approximately $\lambda < 1/(N^2 \alpha)$, which in our case is $\lambda < 40$. Allocating more slots per station brings a negligible improvement in stability, but has a significant effect on performance. What happens is that an increase in k leads to an increase in the number of departing messages from each queue per folder visit, together with a proportionate increase in the intervals between visits. Replacing frequent single departures with less frequent batch departures degrades the performance, by causing messages to remain in the queue for longer.

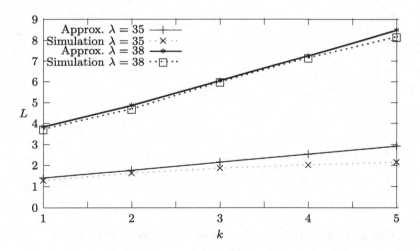

Fig. 2. Symmetric system. Increasing slot allocation

In order to illustrate and quantify the above phenomenon, the allocation of slots to each station was varied between $k = 1$ and $k = 5$. This was done for two different arrival rates, $\lambda = 35$ and $\lambda = 38$, and in each case the approximated value of the average queue size was compared with the simulated one. The other parameters are te same as in te first example. The resulting plots are displayed in Fig. 2.

In this experiment, the approximate values of L were computed by applying the fixed-point iterations (15), with state-dependent departure rates given by (16). The resulting plots suggest that the average queue sizes increase almost linearly with the slot allocation. Moreover, the gradient of that increase is higher when the offered load is higher. The accuracy of the approximations as compared with simulations is again good, especially under the higher load. In the more lightly loaded system, the discrepancy between approximations and simulations appears to grow with k.

In a non-symmetrical setup, the situation can be very different. Our next experiment involves a system where station 1 is much more important than the other stations. The objective is to minimise L_1, the average length of Q_1, regardless of what happens at the other queues. In this example, stations $2, 3, \ldots, N$ are allocated 1 slot each, while k_1, the station 1 allocation, is varied. Moreover, queues $2, 3, \ldots, N$ are assumed to be never empty, i.e. each of their occupancies is $s = 1$. Thus, the environment of Q_1 is defined by the state-dependent departure rates (16), which now have the form

$$\mu_j = [N(j + N - 1)\alpha + \beta]^{-1} \ ; \ j = 1, 2, \ldots, k_1 \ . \tag{22}$$

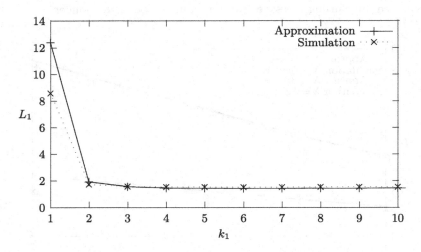

Fig. 3. Asymmetric system. Increasing slot allocation for station 1

In Fig. 3 the approximated and simulated values of L_1 are plotted against the slot allocation k_1. The parameters α and β are the same as before, while the Q_1 arrival rate is fixed at $\lambda_1 = 37$. That constitutes a heavy load when $k_1 = 1$ (remember that then we must have $\lambda_1 < 40$ for stability), but the load decreases when k_1 increases. The resulting average queue length is computed according to (12), without the need for fixed-point iterations.

We observe that the largest performance benefit is derived by increasing the slot allocation from $k_1 = 1$ to $k_1 = 2$. The reduction in the average queue size is substantial. However, further increases in k_1 have a very limited effect because the advantage of larger departing batches is counterbalanced by the larger intervals between departures.

The next and final experiment aims to evaluate the effect of different distributions of folder processing times on performance. This was done entirely by simulation. Remember that when a total of S slots in the folder are occupied, the average interval until it reaches the next station is $v = S\alpha + \beta/N$, and the simulation generates a random number with mean v. Now the distribution of that random number is varied. Five different distributions with the same mean but increasing coefficients of variation were tried.

1. Constant: the interval is equal to v. The coefficient of variation is $C^2 = 0$.
2. Uniform: the interval is distributed uniformly between 0 and $2v$. The coefficient of variation is $C^2 = 1/3$.
3. Exponential: the interval is distributed exponentially with mean v (this has been the case so far). The coefficient of variation is $C^2 = 1$.
4. Hyperexponential A: with probability $2/3$ the interval is distributed exponentially with mean $v/2$ and with probability $1/3$ it is distributed exponentially with mean $2v$. The coefficient of variation is $C^2 = 2$.
5. Hyperexponential B: with probability $5/6$ the interval is distributed exponentially with mean $v/5$ and with probability $1/6$ it is distributed exponentially with mean $5v$. The coefficient of variation is $C^2 = 37$.

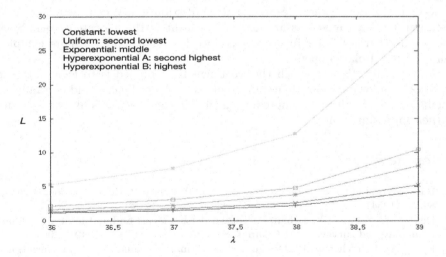

Fig. 4. Different distributions of processing times

The other parameters are the same as in the first example. The arrival rate is increased and the average size of the isolated queue, L, is plotted against λ. The results are displayed in Fig. 4.

It is well known that the performance of any queueing system tends to deteriorate when the coefficient of variation of either the arrival process or the service process increases. This phenomenon is clearly illustrated here. Each increase in C^2 causes the corresponding plot to be higher than the one before. Moreover, the performance deterioration becomes worse as the offered load increases.

Note that for the first four distributions, $C^2 \leq 2$. Within that range, the approximation based on the exponential distribution would be quite acceptable, even at heavy loads. However, the coefficient of variation of the Hyperexponential B distribution is an order of magnitude higher and the corresponding performance is worse by a factor of 3. Then the exponential approximation would not be acceptable. On the other hand, it is unlikely that loading and copying times would have such a high coefficient of variation in practice.

5 Conclusion

We have proposed, analysed and evaluated a new protocol for ordering messages in a distributed environment. The solution enabling the computation of performance measures is approximate but its accuracy appears to be good. This has been verified by comparisons with simulations.

The assumption that one message fits in one slot can be relaxed considerably, while applying the developed general approximation methodology. Each station could be allocated a storage block of a given size in the folder, and message sizes could be i.i.d. random variables with a given distribution. When the folder visits a station, as many messages are loaded as can fit in the allocated block. Such a generalisation is left for future work, since it would require a more complex analysis of an isolated queue.

Another direction in which the work can be extended is to generalize the protocol by allowing more than one circulating folder. That would certainly be worth doing and will be attempted, even though the analysis is likely to require further approximations.

References

1. Bailey, N.T.J.: On queueing processes with bulk service. J. Roy. Stat. Soc. B **16**(1), 80–87 (1954)
2. Défago, X., Schiper, A., Urbán, P.: Total order broadcast and multicast algorithms: taxonomy and survey. ACM Comput. Surv. (CSUR) **36**(4), 372–421 (2004)
3. Ejem, A., Ezhilchevan, P.: Design and performance evaluation of raft variations. In: 39th Annual UK Performance Engineering Workshop (2023)
4. Fouto, P., Preguiça, N., Leitão, J.: High throughput replication with integrated membership management. In: 2022 USENIX Annual Technical Conference (USENIX ATC 22), pp. 575–592 (2022)

5. Guerraoui, R., Ron, R., Pochon, B., Quéma, V.: Throughput optimal total order broadcast for cluster environments. ACM Trans. Comput. Syst. (TOCS) **28**(2), 1–32 (2010)
6. Junqueira, F., Reed, B.: ZooKeeper: Distributed Process Coordination. O'Reilly Media, Inc. (2013)
7. Liskov, B., Cowling, J.: Viewstamped replication revisited (2012)
8. Marandi, P.J., Primi, M., Schiper, N., Pedone, F.: Ring Paxos: a high-throughput atomic broadcast protocol. In: 2010 IEEE/IFIP International Conference on Dependable Systems & Networks (DSN), pp. 527–536 (2010)
9. Oki, B.M., Liskov, B.H.: Viewstamped replication: a new primary copy method to support highly-available distributed systems. In: Proceedings of the 7th Annual ACM Symposium on Principles of Distributed Computing, pp. 8–17 (1988)
10. Ongaro, D., Ousterhout, J.: In search of an understandable consensus algorithm. In: 2014 USENIX Annual Technical Conference (USENIX ATC 14), pp. 305–319 (2014)
11. Schneider, F.B.: Implementing fault-tolerant services using the state machine approach: a tutorial. ACM Comput. Surv. (CSUR) **22**(4), 299–319 (1990)
12. Takagi, H.: Queuing analysis of polling models. ACM Comput. Surv. **20**, 5–28 (1988)
13. Verbitski, A., et al.: Amazon aurora: design considerations for high throughput cloud-native relational databases. In: Proceedings of the 2017 ACM International Conference on Management of Data, pp. 1041–1052 (2017)

The Omnibus Java Library: Efficient Synthesis of Optimal Signal Schedules for Multimodal Intersections

Nicola Bertocci, Laura Carnevali, Leonardo Scommegna[✉], and Enrico Vicario

Department of Information Engineering, University of Florence, Florence, Italy
{nicola.bertocci,laura.carnevali,leonardo.scommegna,
enrico.vicario}@unifi.it

Abstract. The Omnibus Java library efficiently derives optimal signal schedules for multimodal intersections. Specifically, intersections among tram lines with right of way and vehicle flows are considered, minimizing the maximum expected percentage of queued vehicles of each flow.

Trams are modeled by Stochastic Time Petri Nets (STPNs), capturing periodic tram departures and bounded delays and travel times with general (i.e., non-Exponential) distribution. Vehicles are modeled by finite-capacity vacation queues with general vacation times determined by the intersection availability. For each vehicle flow, the expected queue size over time is derived, as well as the steady-state distribution of the expected queue size at multiples of the hyperperiod (resulting from nominal tram arrival times and vehicle traffic signals). Then, the behavior of each vehicle flow can be studied over intervals of arbitrary duration by just performing transient analysis for the hyperperiod duration, starting from the steady-state distribution of the expected queue size.

Omnibus is notably designed to facilitate code usability, maintainability, and extensibility. It is available open source under the AGPLv3 licence. In particular, Omnibus leverages the SIRIO Library of the ORIS tool to model duration distributions and to specify and analyze STPNs.

Keywords: Multimodal intersections · optimal signal schedules · stochastic time Petri nets · finite-capacity vacation queues with general vacation time · Simulation of Urban MObility (SUMO) · software tools and libraries

1 Introduction

1.1 Motivation

The development of tramways is promoted to meet the needs of urban transport while improving its environmental sustainability [1]. The actual success of this initiative depends on the deployment of measures to mitigate the impact of tramways, which in fact reduce the space available for road traffic and typically have right of way at *multimodal* intersections. The main mitigation measure is to optimize traffic signals to minimize the duration of the intervals of intersection unavailability for road transport. To this end, quantitative evaluation of stochastic models capturing behavior of road traffic and tram traffic at multimodal intersections can effectively support early assessment and runtime adaptation of design choices, notably exploiting widespread smart technologies for online estimation of traffic parameters and tram delays [14], so as to maximize expected capacity or minimize expected queue lengths and delays [6]. In addition, a well-engineered implementation of quantitative evaluation methods is also needed to achieve their full exploitation, notably to support the derivation of traffic signal schedules that optimize some performance measure of interest.

1.2 Related Works

A variety of approaches supports operation and management of urban transportation systems, using models with different abstraction level [13]. Notably, *microscopic* models mainly capture behavior of individual vehicles and drivers, while *macroscopic* models capture global or aggregated features of traffic flows, typically achieving computational efficiency while not representing synchronous events, such as tram arrivals. These approaches leverage various modeling formalisms [22,27,28] and solution techniques [11,24,36,40,42] to compute quantitative measures of interests [13], such as the expected queue lengths at intersections. Like the approach implemented by the library presented in this paper, these methods typically do not assume pervasive smart technologies, nor they require significant amounts of mobility data as input. Conversely, other methods optimize traffic signals by leveraging accurate information on vehicle position and movements [17,21,23,31,33], thus requiring advanced vehicle technologies, or by exploiting machine learning methods [2,7,13,32,41], thus requiring availability of large amounts of mobility data. According to this, these latter approaches are considered out of scope with respect to the contribution of this paper.

Specifically, concerning methods based on *microscopic* models, a model of multiple signalized intersections is defined in [9] using Deterministic and Stochastic Petri Nets (DSPNs) [25], approximating normal distributions of vehicle flows and travel times with Erlang distributions, and deriving the durations of traffic light phases that minimize queue lengths. In [4,5], a microscopic model of a road-tramway intersection is presented, modeling periodic tram departures and stochastic tram delays and travel times by Stochastic Time Petri Nets (STPNs) [39], while explicitly representing each state of the queue of vehicles, and thus providing only a rough estimate on the average queue size over time by grouping car arrivals into platoons. Other approaches support rule-based simulation of different signal policies at signalized intersections under different traffic demands [11], the definition of model predictive control policies [30] for connected multimodal signalized intersections between vehicles and bicycles, the development of traffic control policies [18] in the VISSIM microscopic simulation tool [15,37] to accommodate asynchronous priority requests from different modes of transport. Coordination of tram time timetables and signal timing at intersections with road transport is addressed to optimize tram travel times and timetable adherence, by exploiting smart technologies such as automatic vehicle location and advanced control systems [35] or by predetermining signal coordination based on volumes and operational characteristics of road vehicles [20] or by leveraging a simplified model assuming that the same transit signal priority actions have the same effect irrespective of traffic conditions [44]. Cellar automata models have also been used to analyze signal control mechanisms and tram priority policies at vehicle-tram intersections, though typically considering simplified scenarios, e.g. deterministic tram arrival times not subject to stochastic delays [43] and simplified network topology and traffic load [38].

Concerning methods based on *macroscopic* models, Hybrid Petri Nets (HPNs) are used to represent road transport (continuous dynamics) and traffic light signals (discrete dynamics) for both individual [10,12] and multiple intersections [8]. Macroscopic models are also defined by capturing the vehicle flow dynamics at each intersection [45] or by exploiting the input-output approach [34] and the shockwave theory [36]. As a common trait, these approaches typically model the vehicle flow dynamics and the temporization of traffic signal schedules, while not capturing tram traffic behavior characterized by periodic departure times.

To the best of our knowledge, none of the reviewed papers provides an open-source implementation supporting accurate and efficient derivation of optimal signal schedules at multimodal intersections among tram lines and vehicle lows.

1.3 Contribution

We present Omnibus,[1] a Java library available open-source under the AGPL v3 licence, implementing the compositional approach of [3] for efficient derivation of optimal signal schedules for multimodal intersections among tram lines with right of way and vehicle flows. To this end, the approach combines the analysis of a microscopic model of tram traffic specified by STPNs, capturing periodic tram departures as well as bounded delays and travel times with general (non-Exponential) distribution, with the analysis of a macroscopic model of vehicle flows in terms of finite-capacity vacation queues with general vacation times. Notably, the approach of [3] just assumes the presence of sensors detecting tram passages and the availability of statistics of inter-arrival and travel times of vehicles and trams. Omnibus exploits the SIRIO Library[2] of the ORIS tool [29][3] to represent Probability Density Functions (PDFs) of durations as well as to model and evaluate STPNs. It is designed to facilitate code usability, maintainability, and extensibility, by exploiting consolidated design patterns [16].

In the following, first we provide an overview on the Omnibus features, also presenting the typical workflow (Sect. 2). Then, we illustrate the Omnibus packages that support modeling of multimodal intersections (Sect. 3) and evaluation of the expected queue size over time of each vehicle flow (Sect. 4), and we discuss how the obtained results can be exploited to derive optimal signal schedules

---

[1] https://doi.org/10.5281/zenodo.13907731.
[2] https://github.com/oris-tool/sirio.
[3] https://www.oris-tool.org.

Efficient Synthesis of Optimal Signal Schedules for Multimodal Intersections 17

(Sect. 5). Finally, we draw our conclusions and we discuss possible future extensions of the library (Sect. 6).

2 The Omnibus Library

The use case diagram of Fig. 1a shows the main functionalities, implemented by the packages shown in the package diagram of Fig. 1b, i.e., tram, vehicle, and intersection. Figure 1c shows the data flow diagram of the typical workflow. Specifically, first the intersection is defined; then, the STPNs modeling tram tracks are generated and analyzed to derive the probability that the intersection is available for vehicles; next, the queues modeling vehicle flows are generated and analyzed to derive the transient probability of each queue state after a

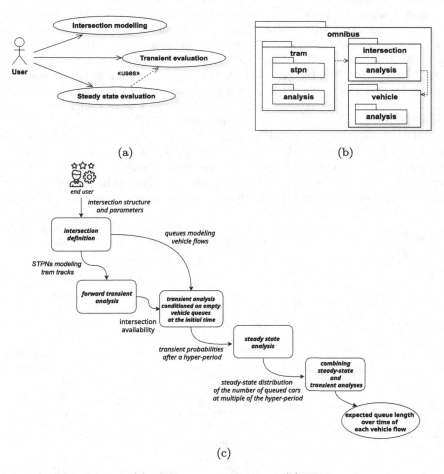

Fig. 1. Omnibus library: (a) UML use case diagram, (b) UML package diagram, and (c) UML data flow diagram of the typical modeling and evaluation workflow.

hyperperiod (i.e., least common multiple of periods of traffic signals and tram lines); finally, steady-state and transient analyses of vehicle flows are combined to derive the expected queue length over time over intervals of arbitrary duration.

3 Intersection Modeling

Tram and Vehicle Traffic Model. A tram track is modeled by an STPN, where deterministic (DET) transitions represent periodic tram departures and offsets, while general (GEN) transitions model tram delays and travel times. Each vehicle flow is modeled by a finite-capacity vacation queue with EXP (Exponential) inter-arrival times and service times, and GEN vacation times determined by the probability that the intersection is available for vehicles and by the time-division multiplexing schedule of the intersection traffic light.

Figure 2 shows a multimodal intersection considered in [3], whose main stochastic parameters are reported in Table 1. Specifically, the intersection consists of a bidirectional tram line, i.e., a tram line made of two tracks ϕ_{11}^{tram} and ϕ_{12}^{tram}, and three vehicle flows ϕ_1^{veh}, ϕ_2^{veh}, and ϕ_3^{veh}. Trams of both tracks have arrival period $T = 220$ s, travel time from the wayside system to the intersection equal to 5 s (a wayside system is installed on each track, close to the intersection, to detect tram passages and trigger the traffic signals of vehicles red as the tram is approaching), and crossing time uniformly distributed over $[6, 14]$ s. The offset of tram arrivals (i.e., nominal arrival time of the first tram of a track) is equal to 0 s for the first track and equal to 40 s for the second track, and the delay with respect to the nominal arrival time at the wayside system is uniformly distributed over $[0, 120]$ s for the first track and $[0, 40]$ s for the second one. It is worth noting that these values of the intersection parameters tend to synchronize

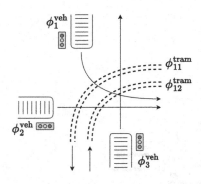

Fig. 2. Multimodal intersection considered in [3], with 2 tram tracks and 3 vehicle flows.

Table 1. Main stochastic parameters of the tram lines and the vehicle flows of the multimodal intersection represented in Fig. 2.

parameter	value
tram line period T	220 s
Φ_{11}^{tram} offset	0 s
Φ_{12}^{tram} offset	40 s
Φ_{11}^{tram} delay distribution	UNIF(0 s, 120 s)
Φ_{12}^{tram} delay distribution	UNIF(0 s, 40 s)
$\Phi_{11}^{\text{tram}}, \Phi_{12}^{\text{tram}}$ red signal trigger time	5 s
$\Phi_{11}^{\text{tram}}, \Phi_{12}^{\text{tram}}$ crossing time distribution	UNIF(6 s, 14 s)
traffic light period P	110 s
Φ_1^{veh} arrival rate	0.05 s^{-1}
Φ_2^{veh} arrival rate	0.1 s^{-1}
Φ_3^{veh} arrival rate	0.15 s^{-1}
$\Phi_1^{\text{veh}}, \Phi_2^{\text{veh}}, \Phi_3^{\text{veh}}$ leaving rate	0.092 s^{-1}
$\Phi_1^{\text{veh}}, \Phi_2^{\text{veh}}, \Phi_3^{\text{veh}}$ queue capacity	31

tram passages, and thus to increase the duration of the intervals of intersection unavailability for vehicles, thus making the search for optimal schedules more challenging.

Concerning vehicle flows, the traffic light period P is equal to 110 s, i.e., half of the tram departure period. Vehicle flows Φ_1^{veh}, Φ_2^{veh}, and Φ_3^{veh} have arrival rate equal to $0.05\,\text{s}^{-1}$, $0.1\,\text{s}^{-1}$, and $0.15\,\text{s}^{-1}$, respectively. All vehicle flows have leaving rate equal to $0.092\,\text{s}^{-1}$, which results from street length equal to 150 m and maximum vehicle speed equal to $50\,\text{km}\,\text{h}^{-1}$ (the latter two parameters are not shown in Table 1). Moreover, all vehicle flows have queue capacity $K = 31$, which results from the mentioned street length and from vehicle length equal to 4.5 m and safe distance between vehicles equal to 0.3 m) (also in this case, the latter two parameters are not shown in Table 1). We remark that street length, vehicle length, and safe distance between vehicles are not parameters of the queues modeling vehicle flows. Rather, they are typical parameters of microscopic traffic simulators, like SUMO (Simulation of Urban MObility) [26]. Therefore, such parameters are considered to facilitate the comparison of the

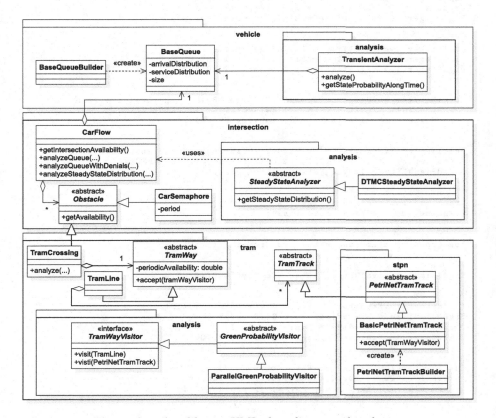

Fig. 3. Omnibus library: UML class diagram of packages.

experimental results with those achieved by using tools like SUMO. This kind of validation is in fact performed in [3] for the optimization of signal schedules.

Implementation. Figure 3 shows the Omnibus class diagram. Each vehicle flow (class `CarFlow`) is associated with a queue (class `BaseQueue`) and a set of obstacles (class `Obstacle`), i.e., semaphores (class `CarSemaphore`) and tram crossings (class `TramCrossing`) related to tram lines (class `TramLine`) composed of tracks (class `TramTrack`). Omnibus emphasizes the concept of vehicle flow rather than that of intersection, reflecting the implemented compositional approach.

Listing 1.1 constructs the model of the intersection shown in Fig. 2.

```
1  // tram line parameter definition
2  BigInteger tramPeriod = BigInteger.valueOf(220);
3
4  // track 1 parameter definition
5  BigInteger t1_offsetTime = BigInteger.ZERO;
6  BigInteger t1_delayEFTime = BigInteger.ZERO;
7  BigInteger t1_delayLFTime = BigInteger.valueOf(120);
8
9  // track 2 parameter definition
10 BigInteger t2_offsetTime = BigInteger.valueOf(40);
11 BigInteger t2_delayEFTime = BigInteger.ZERO;
12 BigInteger t2_delayLFTime = BigInteger.valueOf(40);
13
14 // track 1 and track 2 parameter definition
15 BigInteger redSignalTime = BigInteger.valueOf(5);
16 BigInteger leavingEFTime = BigInteger.valueOf(6);
17 BigInteger leavingLFTime = BigInteger.valueOf(14);
18
19 // track 1 instantiation
20 PetriNetTramTrack bin1 = PetriNetTramTrackBuilder.getInstance("bin1",
21    tramPeriod, t1_offsetTime, t1_delayEFTime, t1_delayLFTime,
22    redSignalTime, leavingEFTime, leavingLFTime);
23
24 // track 2 instantiation
25 PetriNetTramTrack bin2 = PetriNetTramTrackBuilder.getInstance("bin2",
26    tramPeriod, t2_offsetTime, t2_delayEFTime, t2_delayLFTime,
27    redSignalTime, leavingEFTime, leavingLFTime);
28
29 // tram line definition
30 TramLine tramLine = new TramLine("line1");
31 tramLine.addTramTrack(bin1, bin2);
32
33 // tram cross definition
34 TramCrossing tramCross = new TramCrossing(tramLine);
35
36 // vehicle flow 1 parameter definition
37 BigDecimal arrivalRate1 = BigDecimal.valueOf(0.05);
38
39 // vehicle flow 1, vehicle flow 2, and vehicle flow 3 parameter definition
40 BigInteger carSemaphorePeriod = BigInteger.valueOf(110);
41 BigDecimal mu = BigDecimal.valueOf(0.092);
42 BigInteger maxQueueSize = BigInteger.valueOf(31);
43 BigInteger initialCars = BigInteger.valueOf(0);
44
45 // vehicle flow 1 instantiation
46 CarFlow carFlow1 = new CarFlow("carFlow1");
47 carFlow1.setQueue(BaseQueueBuilder.getInstance(
48    arrivalRate1, mu, maxQueueSize, initialCars));
49
50 // vehicle flow 2 and 3 instantiation
51 ...
52
```

```
53 // semaphore instatiation with period 110s
54 CarSemaphore carSem1 = new CarSemaphore(carSemaphorePeriod, TIMESTEP);
55 CarSemaphore carSem2 = new CarSemaphore(carSemaphorePeriod, TIMESTEP);
56 CarSemaphore carSem3 = new CarSemaphore(carSemaphorePeriod, TIMESTEP);
57
58 // adding obstacles to vehicle flow 1
59 carFlow1.addObstacle(tramCross);
60 carFlow1.addObstacle(carSem1);
61
62 // adding obstacle to vehicle flows 2 and 3
63 ...
```

Listing 1.1. Construction of the model of the intersection of Fig. 2.

4 Intersection Evaluation

Tram and Vehicle Traffic Analysis. The STPN of each tram track is analyzed in isolation by forward transient analysis based on the method of stochastic state classes [19], deriving the transient probability that the intersection is available for each vehicle flow. For every flow, a set of ordinary differential equations is defined, where the leaving rate of vehicles is modulated by the probability that the intersection is available for their transit. For each vehicle flow, the solution of this set of equations yields the expected queue size over time.

Within a few hyperperiods, the distribution of the expected queue size of each vehicle flow reaches a steady state at multiples of the hyperperiod. In particular, the distribution of the expected queue size of each vehicle flow at multiples of the hyperperiod can be derived by steady-state analysis of the Discrete Time Markov Chain (DTMC) embedded in the continuous-time birth-death process of the queue at multiples of the hyperperiod, performing transient analysis within a hyperperiod to derive the DTMC transition probability matrix. Therefore, once the mentioned steady-state distribution is derived for each vehicle flow, the evaluation of the expected queue size over time can be performed over time-intervals of arbitrary duration, even in presence of time-varying stochastic parameters. It is worth remarking that deriving such steady-state distribution by using a microscopic traffic simulator would be almost unfeasible, as discussed in [3]. In fact, the analysis time using Omnibus is up to nearly four orders of magnitude lower than the simulation time using SUMO, notably evaluating in few minutes hundreds of schedules requiring tens of hours in SUMO, which actually prevents using SUMO when a significant number of simulation runs is required, as in the case of the evaluation of the mentioned steady-state distribution.

Implementation. The probability that the intersection is available for vehicles is derived by class `ParallelGreenVisitor`, implemented using the Visitor design pattern [16], allowing the analysis to be performed both for single tracks (class `BasicPetriNetTramTrack`) and for all tracks (class `TramWay`). For each vehicle flow, transient and steady-state analysis are performed by classes `TransientAnalyzer` and `SteadyStateAnalyzer`, respectively. Listing 1.2 illustrates these concepts by analyzing the model built in Listing 1.1.

```
1  // tram crossings analysis
2  tramCross.analyze(new ParallelGreenProbabilityVisitor(), timeStep);
3
4  // transient analysis
5  double[] expectedState = carFlow.analyzeQueue(new TransientAnalyzer(),
       BigInteger.valueOf(carFlow.getObstaclesHyperPeriod()), timeStep).
       getExpectedStateAlongTime());
6
7  // steady-state analysis
8  double[] steadyStateDistribution = carFlow1.analyzeSteadyStateDistribution
       (new DTMCSteadyStateAnalyzer(), new TransientAnalyzer(), timeStep).
       getSteadyStateDistribution();
```

Listing 1.2. Evaluation of the intersection model built in Listing 1.1.

5 Efficient Derivation of Optimal Signal Schedules

Defining and Exploring a Set of Static Signal Schedules. A set of static traffic signal schedules can be defined starting by varying sequence and duration of phases. In particular, in [3], a set of 390 different traffic signal schedules is defined as follows. During a traffic light period $P = 110$ s, phases of duration $\Delta \in \{15, 25, 35\}$ s are considered, except for the last phase which may be longer. A short time interval of 5 s during which all vehicle signals are red is considered between any two phases assigned to vehicle flows. Finally, to guarantee that each phase has duration at least equal to 15 s, the time until the end of the period is assigned to the second-to-last phase if the last phase lasted less than 15 s. For instance, within each period $[0, 110]$ s, a possible schedule is the one that assigns the time interval $[0, 25]$ s to the first vehicle flow ϕ_1^{veh}, the time interval $[30, 65]$ s to the second vehicle flow ϕ_2^{veh}, and the time interval $[70, 105]$ s to the third vehicle flow ϕ_3^{veh}. During the intervals $(25, 30)$ s and $(65, 70)$ s the signal is red for all the vehicle flows. Also note that, when a time interval is assigned to a vehicle flow, then the signal during that time interval is green for the vehicle flow unless a tram is approaching or crossing the intersection.

For each schedule, the analysis is performed up to 5 hyperperiods (i.e., 110 s) with time step equal to 0.1 s, For each schedule, 25 simulation runs are also performed through the SUMO microscopic traffic simulator. In both cases, the computed measure of interest is the maximum expected percentage of queued vehicles of any flow within the interval comprised between 2 and 5 hyperperiods. The optimality of each of the 390 traffic signal schedule is evaluated in terms of minimizing this quantitative measure by using SUMO, and the obtained results are compared with those achieved by using Omnibus.

Figure 4 shows the expected queue size over time of each vehicle flow starting from its steady-state distribution at multiples of the hyperperiod, computed for the best schedule, the worst schedule, and the schedule in median position, computed by using SUMO. Note that the best and worst schedules can be identified also using Omnibus, notably with a significantly lower computation time.

Implementation. Listing 1.3 provides a proof of concept demonstrating how Omnibus can be used to efficiently derive optimal schedules. The

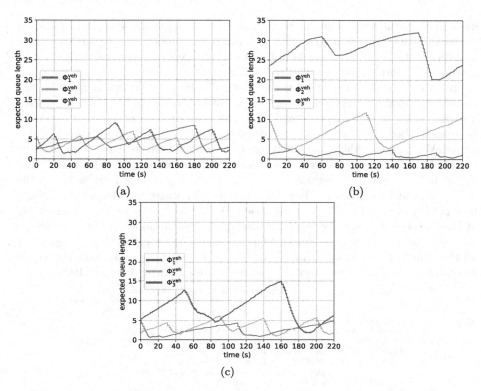

Fig. 4. For each vehicle flow of the intersection of Fig. 2, expected queue size over time starting from the steady-state distribution at multiples of the hyperperiod, obtained in [3] for (a) the best schedule, (b) the worst schedule, and (c) the schedule in median position among the 390 considered schedules.

snippet is divided into two methods, i.e., `identifyOptimalSchedule` and `getMaxOccupationPer-centage`, thus enhancing readability and flexibility. On the one hand, `identify-OptimalSchedule` handles the initial setup and the iteration over the schedules to be analyzed. On the other hand, `getMaxOccupationPercentage` is responsible for analyzing how a schedule affects a vehicle flow by calculating the maximum percentage of queue occupation at multiples of the hyperperiod. There is a functional dependency between the two methods: `getMaxOccupationPercentage` will be invoked by `identifyOptimalSchedule` to analyze each vehicle flow at the intersection for each schedule. In turn, `identifyOptimalSchedule` will be invoked by the end-user in order to obtain the best schedule.

Note that the best schedule is currently identified as the one that minimizes the maximum percentage of queue occupation in any flow of the intersection. However, other criteria could be considered as well, depending on the specific context and aim of the optimization. For example, in case some queues might be more important than others, and thus a weighted optimization should be implemented. Alternatively, it might be desirable to optimize the absolute number of vehicles queued at the intersection instead of the percentage. In such cases, it is sufficient to replace the method `getMaxOccupationPercentage` with the new method that implements the desired heuristic.

In more detail, starting from the method `identifyOptimalSchedule`, the method `generateSemaphoreSchedules` is invoked in line 5. This method generates the space of schedules to explore in the form of Java Strings and it is not represented because it is considered trivial and not crucial for the purposes of this paper. We just provide some details on the format of the schedule. Specifically, the id of the vehicle flow for which the semaphore is exclusively green is repeated for each time unit. For example, if the id of the object `carFlow1` is 1 and the id of the object `carFlow2` object is 2, then a valid schedule with period 6 could be "111222" and "112211". As discussed in [3] and in the previous paragraph of this section, it could be realistic considering semaphore schedules in which the right of way of vehicle flows is intertwined with intervals of red signal for all the vehicle flows. In this case, using id 9 to identify time intervals, the above mentioned schedule would become the one represented by String "119229".

```java
public String identifyOptimalSchedule(CarFlow[] carFlows,
                                      CarSemaphore[] carSemaphores)
{
    String bestSchedule;
    double bestScheduleOccupationPercentage = Double.MAX_VALUE;
    BlockingQueue schedulesToAnalyze = generateSemaphoreSchedules();
    for (String schedule: schedulesToAnalyze) {
        Utils.applySchedule(schedule, carSemaphores);
        double maxOccupationPercentage = Double.MIN_VALUE;
        for (CarFlow carFlow: carFlows) {
            double w = getMaxOccupationPercentage(carFlow);
            maxOccupationPercentage = Math.max(maxOccupationPercentage,w);
        }
        if (maxOccupationPercentage < bestScheduleOccupationPercentage) {
            bestSchedule = schedule;
            bestScheduleOccupationPercentage = maxOccupationPercentage;
        }
    }
    return bestSchedule;
}

private double getMaxOccupationPercentage(CarFlow carFlow) {
    double[] steadyStateDistribution = carFlow
        .analyzeSteadyStateDistribution(
        new DTMCSteadyStateAnalyzer(),
        new TransientAnalyzer(),
        timeStep).getSteadyStateDistribution();

    BigDecimal[] bdDist = Arrays.stream(steadyStateDistribution)
        .mapToObj(BigDecimal::valueOf)
        .collect(Collectors.toList())
        .toArray(new BigDecimal[0]);

    carFlow.getQueue().setInitialDistribution(bdDist);

    return Arrays.stream(carFlow.analyzeQueue(new TransientAnalyzer(),
        BigInteger.valueOf(carFlow.getObstaclesHyperPeriod()),timeStep).
        getExpectedStateAlongTime()).max().getAsDouble()/carFlow.getQueue().
        getSize().doubleValue();
}
```

Listing 1.3. Implementation of the optimal signal schedule identification.

Once the schedule space is generated, it is explored using the for loop in line 6. In line 7, the method `applySchedule` is invoked. This is a utility function that sets the green and red intervals for each semaphore according to the schedule under consideration. Subsequently, for each vehicle flow at the intersection, the method `getMaxOccupationPercentage` is invoked, calculating the maximum expected percentage of queued vehicles of each flow. For each schedule, the maximum expected percentage of queued vehicles among all the flows of the intersection is selected (lines 10 and 11). Once this measure of interest (i.e., `maxOccupationPercentage`) is identified, the schedule that yields the smallest value of this measure is selected as the best schedule (lines 13 to 15).

The core of the process however is the method `GetMaxOccupationPercentage` (line 21). Given a vehicle flow, the method returns the maximum expected percentage of queued vehicles during a hyperperiod starting from thes steady-state distribution (of the number of queued vehicles at multiples of the hyperperiod). This calculation is made possible by transient and steady-state analyses provided by the Omnibus Java Library. The mentioned steady-state distribution is calculated for a vehicle flow at line 22, equivalently to how it was illustrated in Listing 1.2 (line 8). Subsequently, the queue occupation is set equal the value obtained at the steady state in the previous step (line 33). Finally, transient analysis is performed (line 35), again equivalently to how it was illustrated in Listing 1.2 (line 5). In the same instruction, the maximum expected value in the hyperperiod is extracted and the percentage of occupation is calculated (line 41).

6 Conclusions and Future Extensions

We have presented the Omnibus Java library, implementing efficient derivation of optimal signal schedules for multimodal intersections [3]. Specifically, intersections among tram lines with right of way and vehicle flows are considered in Omnibus, combining a microscopic model of tram traffic (i.e., STPNs representing periodic tram departures, and bounded delays and travel times with non-Exponential distribution) with a macroscopic model of vehicle flows (i.e., finite-capacity vacation queues with general vacation times). The analyses of both models are also combined in [3], developing an efficient compositional approach to compare the intersection performance (in terms of expected percentage of queued vehicles of each flow) under different signal schedules. Therefore, the implementation of the approach provided by Omnibus can be effectively used to analyze a significant number of signal schedules and derive an optimal solution.

The Omnibus library is specifically designed to facilitate code usability, maintainability, and extensibility, supporting a variety of extensions that also comprise relevant advancements in the theoretical and application perspective. In particular, Omnibus could be easily extended to represent vehicle arrival times in the class of time-inhomogeneous Poisson Processes, enabling representation of vehicles arriving in bursts, platoons, and free flow. Other non-Exponential distributions that can be represented by a Markovian model, such as hyper-Exponential distributions and Markovian Arrival Processes, could be implemented as well to represent arrival times and leaving times of vehicles. In both cases, though the advancement involves both the modeling and evaluation perspectives, the extension of the implementation remains mainly confined to a couple of classes. Namely, `BaseQueue`, possibly implementing a subclass, and `TransientAnalyzer`, adapting the analysis to the arrival and service time distributions of the new subclass, possibly by implementing a strategy method [16].

Another possible extension that would bring greater expressiveness to the Omnibus Java Library could consist in considering pedestrian crossing as an additional factor for the intersection. This would exclusively require the implementation of a new subclass of the class `Obstacle` where, similarly to the classes `CarSemaphore` and `TramCrossing`, the designer defines the law according to which pedestrians occupy a lane while crossing the intersection.

Moreover, the Omnibus library could also be extended to represent *multiple* connected intersections within a urban transportation network, enabling joint optimization of traffic signal schedules. Similarly to the previous case, thanks to the flexible architecture of the library, the extension would mainly require the addition of new classes (e.g., to define how the intersections are connected with each other) rather that the modification of the existing ones.

Notably, all these potential extensions demonstrate how Omnibus adheres to the Open/Closed Principle. In fact, these additions would not alter the existing API exposed to end-users (including the methods invoked in Listing 1.1 to 1.3). Rather, they comprise extensions of the API, allowing the designer to instantiate new types of obstacles and specify new types of arrival and crossing patterns without creating relevant incompatibilities with the new versions of the library.

Acknowledgement. This work was partially supported by the European Union under the Italian National Recovery and Resilience Plan (NRRP) of NextGenerationEU, partnership on "Telecommunications of the Future" (PE00000001 - program "RESTART"), and by the MUR PRIN 2022 PNRR P2022A492B project ADVENTURE (ADVancEd iNtegraTed evalUation of Railway systEms) funded by the European Union - NextGenerationEU.

References

1. ACEA: The 2030 urban mobility challenge. Technical report (2016)
2. Balaji, P., Srinivasan, D.: Multi-agent system in urban traffic signal control. IEEE Comput. Intell. Mag. **5**(4), 43–51 (2010)
3. Bertocci, N., Carnevali, L., Scommegna, L., Vicario, E.: Efficient derivation of optimal signal schedules for multimodal intersections. Simul. Modelling Pract. Theory 102912 (2024)
4. Carnevali, L., Fantechi, A., Gori, G., Vicario, E.: Analysis of a road/tramway intersection by the ORIS tool. In: International Conference on Verification and Evaluation of Computer and Communication Systems, pp. 185–199. Springer (2018)
5. Carnevali, L., Fantechi, A., Gori, G., Vicario, E.: Stochastic modeling and analysis of road-tramway intersections. Innovations Syst. Softw. Eng. **16**(2), 215–230 (2020)
6. Cheng, C., Du, Y., Sun, L., Ji, Y.: Review on theoretical delay estimation model for signalized intersections. Transp. Rev. **36**(4), 479–499 (2016)

7. Chu, T., Wang, J., Codecà, L., Li, Z.: Multi-agent deep reinforcement learning for large-scale traffic signal control. IEEE Tran. Intell. Transp. Syst. **21**(3), 1086–1095 (2019)
8. Di Febbraro, A., Giglio, D., Sacco, N.: Urban traffic control structure based on hybrid Petri nets. IEEE Tr. Int. Tran. Sys. **5**(4), 224–237 (2004)
9. Di Febbraro, A., Giglio, D., Sacco, N.: A deterministic and stochastic Petri net model for traffic-responsive signaling control in urban areas. IEEE Trans. Int. Transp. Sys. **17**(2), 510–524 (2016)
10. Di Febbraro, A., Sacco, N.: On modelling urban transportation networks via hybrid Petri nets. Control Eng. Pract. **12**(10), 1225–1239 (2004)
11. Dion, F., Hellinga, B.: A rule-based real-time traffic responsive signal control system with transit priority: application to an isolated intersection. Transp. Res. Part B: Methodol. **36**(4), 325–343 (2002)
12. Dotoli, M., Fanti, M.P., Iacobellis, G.: An urban traffic network model by first order hybrid Petri nets. In: 2008 IEEE International Conference on Systems, Man and Cybernetics, pp. 1929–1934. IEEE (2008)
13. Eom, M., Kim, B.-I.: The traffic signal control problem for intersections: a review. Eur. Transp. Res. Rev. **12**(1), 1–20 (2020). https://doi.org/10.1186/s12544-020-00440-8
14. Faria, R., Brito, L., Baras, K., Silva, J.: Smart mobility: a survey. In: International Conference on IoT for the Global Community, pp. 1–8. IEEE (2017)
15. Fellendorf, M.: VISSIM: A microscopic simulation tool to evaluate actuated signal control including bus priority. In: 64th Institute of Transportation Engineers Annual Meeting, vol. 32, pp. 1–9. Springer (1994)
16. Gamma, E., Helm, R., Johnson, R., Vlissides, J., Patterns, D.: Elements of reusable object-oriented software. Design Patterns (1995)
17. Guo, Q., Li, L., Ban, X.J.: Urban traffic signal control with connected and automated vehicles: a survey. Transp. Res. Part C: Emer. Technol. **101**, 313–334 (2019)
18. He, Q., Head, K.L., Ding, J.: Multi-modal traffic signal control with priority, signal actuation and coordination. Transp. Res. Part C: Emer. Technol. **46**, 65–82 (2014)
19. Horváth, A., Paolieri, M., Ridi, L., Vicario, E.: Transient analysis of non-Markovian models using stochastic state classes. Perform. Eval. **69**(7–8), 315–335 (2012). https://doi.org/10.1016/j.peva.2011.11.002
20. Ji, Y., Tang, Y., Du, Y., Zhang, X.: Coordinated optimization of tram trajectories with arterial signal timing resynchronization. Transp. Res. Part C: Emer. Technol. **99**, 53–66 (2019)
21. Ji, Y., Tang, Y., Shen, Y., Du, Y., Wang, W.: An integrated approach for tram prioritization in signalized corridors. IEEE Trans. Intell. Transp. Syst. **21**(6), 2386–2395 (2019)
22. Li, Y., Sun, D.: Microscopic car-following model for the traffic flow: the state of the art. J. Contr. Theory and Appl. **10**(2), 133–143 (2012)
23. Li, Z., Elefteriadou, L., Ranka, S.: Signal control optimization for automated vehicles at isolated signalized intersections. Transp. Res. Part C: Emer. Technol. **49**, 1–18 (2014)
24. Lighthill, M.J., Whitham, G.B.: On kinematic waves II. A theory of traffic flow on long crowded roads. Proc. Roy. Soc. Lon. Ser. A. Math. Phys. Sci. **229**(1178), 317–345 (1955)
25. Lindemann, C.: Performance modelling with deterministic and stochastic petri nets. ACM Sigmetrics Perform. Eval. Rev. **26**(2), 3 (1998)
26. Lopez, P.A., et al.: Microscopic traffic simulation using SUMO. In: International Conference on Intelligent Transportation Systems, pp. 2575–2582. IEEE (2018)

27. Maerivoet, S., De Moor, B.: Cellular automata models of road traffic. Phys. Rep. **419**(1), 1–64 (2005)
28. Ng, K.M., Reaz, M.B.I., Ali, M.A.M.: A review on the applications of Petri nets in modeling, analysis, and control of urban traffic. IEEE Trans. Int. Transp. Syst. **14**(2), 858–870 (2013). https://doi.org/10.1109/TITS.2013.2246153
29. Paolieri, M., Biagi, M., Carnevali, L., Vicario, E.: The ORIS tool: quantitative evaluation of non-Markovian systems. IEEE Trans. Softw. Eng. **47**(6), 1211–1225 (2021)
30. Portilla, C., Valencia, F., Espinosa, J., Nunez, A., De Schutter, B.: Model-based predictive control for bicycling in urban intersections. Transp. Res. Part C: Emer. Technol. **70**, 27–41 (2016)
31. Pourmehrab, M., Elefteriadou, L., Ranka, S., Martin-Gasulla, M.: Optimizing signalized intersections performance under conventional and automated vehicles traffic. IEEE Trans. Intell. Transp. Syst. **21**(7), 2864–2873 (2019)
32. Prabuchandran, K., AN, H.K., Bhatnagar, S.: Multi-agent reinforcement learning for traffic signal control. In: International IEEE Conference on Intelligent Transportation Systems, pp. 2529–2534. IEEE (2014)
33. Reddy, R., Almeida, L., Gaitán, M.G., Santos, P.M., Tovar, E.: Synchronous management of mixed traffic at signalized intersections towards sustainable road transportation. IEEE Access (2023)
34. Sharma, A., Bullock, D.M., Bonneson, J.A.: Input-output and hybrid techniques for real-time prediction of delay and maximum queue length at signalized intersections. Transp. Res. Record **2035**(1), 69–80 (2007)
35. Shi, J., Sun, Y., Schonfeld, P., Qi, J.: Joint optimization of tram timetables and signal timing adjustments at intersections. Transp. Res. Part C: Emer. Technol. **83**, 104–119 (2017)
36. Stephanopoulos, G., Michalopoulos, P.G., Stephanopoulos, G.: Modelling and analysis of traffic queue dynamics at signalized intersections. Transp. Res. Part A: General **13**(5), 295–307 (1979)
37. Stevanovic, J., Stevanovic, A., Martin, P.T., Bauer, T.: Stochastic optimization of traffic control and transit priority settings in VISSIM. Transp. Res. Part C: Emer. Technol. **16**(3), 332–349 (2008)
38. Tonguz, O.K., Viriyasitavat, W., Bai, F.: Modeling urban traffic: a cellular automata approach. IEEE Comm. Maga. **47**(5), 142–150 (2009)
39. Vicario, E., Sassoli, L., Carnevali, L.: Using stochastic state classes in quantitative evaluation of dense-time reactive systems. IEEE Trans. Softw. Eng. **35**(5), 703–719 (2009)
40. Webster, F.V.: Traffic signal settings. Road Research Technical Paper 39. Technical report (1958)
41. Wei, H., Zheng, G., Gayah, V., Li, Z.: A survey on traffic signal control methods. arXiv preprint arXiv:1904.08117 (2019)
42. Yagar, S., Han, B., Greenough, J.: Real-time signal control for mixed traffic and transit based on priority rules. In: Traffic Management. Proceedings of the Engineering Foundation Conference (1992)
43. Zhang, L., Garoni, T.: A comparison of tram priority at signalized intersections. arXiv preprint arXiv:1311.3590 (2013)
44. Zhang, T., Mao, B., Xu, Q., Feng, J.: Timetable optimization for a two-way tram line with an active signal priority strategy. IEEE Access **7**, 176896–176911 (2019)
45. Zhang, Y., Su, R.: An optimization model and traffic light control scheme for heterogeneous traffic systems. Transp. Res. Part C: Emer. Technol. **124**, 102911 (2021)

Performance Evaluation of Beaconing Schemes for Vehicular Platooning

Hassan Laghbi[1,2](✉) and Nigel Thomas[1](✉)

[1] Newcastle University, Newcastle upon Tyne NE1 7RU, UK
{h.laghbi2,nigel.thomas}@ncl.ac.uk
[2] Jazan University, Jazan 82817, Saudi Arabia

Abstract. In this study, we propose PlatoonB, a distributed scheme designed to reduce communication channel load and maintain safety in vehicular platooning networks. Because all the evaluated schemes exhibited their poorest performance in maintaining the minimum inter-vehicle distance during the emergency braking experiment, we further propose PlatoonBE, an enhanced version of PlatoonB which performs better in the braking experiment. We conducted extensive simulations to compare our scheme with five other approaches -Static, Slotted, DynB1, DynB2 and JerkB- focusing on key performance indicators: channel busy ratio, packet collision rate per second, and minimum distance between platoon members. The results reveal that our scheme is light on the communication channel and safer in the braking scenario.

Keywords: V2V · Platooning · Congestion · Beaconing

1 Introduction

Traffic jams, shock waves and high fuel consumption by heavy-duty vehicles can have significant environmental and economic impacts including decreased productivity, more fuel expenses, and higher carbon emissions. An application that aims to mitigate these impacts is vehicular platooning where vehicles move in harmony as a single entity. By reducing the gaps between vehicles, this application not only increases road capacity but also reduces fuel consumption by reducing the air resistance that platoon members face [1]. Vehicular platooning relies on cooperative adaptive cruise control (CACC) systems installed in the vehicles [17]. These controllers need data from other platoon members, including the lead vehicle, data that cannot be fully gathered through sensors alone. Thus, platooning requires a robust vehicle-to-vehicle (V2V) communication network, where vehicles continuously broadcast messages (beacons) containing necessary information including position, speed, and acceleration.

Deciding on optimal beaconing rates poses a challenge, as no single beaconing frequency is universally effective across all scenarios. Static beaconing at a high frequency, for example, 10 or 20 messages per second, can cause channel congestion and lead to frequent packet losses, particularly in situations where there is a high number of vehicles within range. Applications other than platooning, such as cooperative perception [18] and danger warnings, may be also running on the vehicles and require their own share of the communication channel. Conversely, static beaconing at a lower frequency, such as 1 or 2 times per second, may not provide platooning controllers with the timely data necessary to calculate the next input which is crucial for maintaining platoon stability and safety. Dynamic beaconing approaches, such as DCC, LIMERIC and DynB [3,15,16], adapt the beaconing interval to reduce the channel load but without considering platooning requirements which leads to unsafe platooning. DCC and LIMERIC increase or decrease beaconing based on the current CBR (channel busy ratio), while DynB considers both the CBR and the number of neighbouring vehicles within range. However, none of these schemes react to vehicle dynamics making them unsuitable for critical applications like platooning. For example, when a platoon experiences changes in speed and updates need to be exchanged between platoon members, DynB restricts the frequency of updates if the number of neighbours within range is high, which can lead to vehicle crashes. Synchronisation approaches, such as using Time Division Multiple Access (TDMA), like Slotted, or tokens, are effective against packet loss but do not reduce the channel load. Another scheme, Jerk Beaconing [6], does consider vehicle dynamics but is insufficiently reactive in some critical situations such as hard braking in high vehicle density conditions. Our approach reduces the load on the communication channel while remaining highly reactive in such critical scenarios.

The main contributions of this work are as follows:

- We propose PlatoonB, a beaconing scheme for platooning that is light on the communication channel.
- We propose the addition of intent beacons to increase reactivity in the stopping scenario, and implement this within PlatoonB (referred to as PlatoonBE). In the experiments, the earlier version, PlatoonB, is also included to provide a clear comparison and highlight the improvements brought by PlatoonBE.
- We limit the maximum beaconing interval by TTC, time to crash, which dynamically allows vehicles to beacon more when inter-vehicle distance becomes shorter. Calculating this dynamic metric adds no extra overhead since it uses local data from vehicle sensors or information already included in the exchanged beacons.
- We perform extensive simulation experiments to compare the performance of our scheme against other approaches.

The rest of this paper is organised as follows: Sect. 2 reviews the current beaconing schemes. Section 3 presents the considered system model and explains the proposed schemes. Section 4 presents the experiments and results. Section 5 concludes the paper and presents some future directions.

2 Related Work

In [14], Martijn et al. studied the periodic beacon generation rate required for CACC systems, noting that it typically ranges between 10 and 25 Hz. Such Static beaconing is acceptable when the number of vehicles in a range is small and when vehicles are not generating additional traffic from other VANET applications besides platooning. Otherwise, the communication channel could become congested which leads to delays and packet loss.

Distributed Congestion Control (DCC) [15] and LIMERIC [16], are two congestion control algorithms which both rely on the CBR as a key metric for detecting and managing congestion. A commonly used variant of DCC uses a table that matches CBR values to different beaconing intervals. For example, a beaconing interval of 0.5 s is selected when the CBR is greater than 59%. On the other hand, LIMERIC allows each vehicle to adapt its beaconing interval so that the collective CBR eventually reaches a desired value. Sommer et al. [3] designed a dynamic beaconing scheme, DynB, for vehicular ad-hoc networks which adjusts the beaconing interval, I, based on two parameters: b_t a measure of the channel busy time, and N, a count of one-hop neighbours. DynB aims to keep I close to a desired value I_{des} without letting the channel load exceed b_{des}. The beaconing interval is then calculated as $I = I_{des}(1 + rN)$ where r is computed as $r = \frac{b_t}{b_{des}} - 1$. Therefore, the beaconing interval increases with the number of neighbours only when the CBR is above b_{des}.

Segata et al. [4] proposed Slotted Beaconing which is an intra-platoon TDMA approach based on the positions of the platoon members. This approach keeps the beaconing interval static and aims at containing congestion by reducing channel contention. It evenly divides the beaconing interval into fixed time slots for each vehicle's data transmission, starting with the leader and ending with vehicle $n - 1$, in an n-vehicle platoon. Their results show reduction in channel contention among platoon members. A similar TDMA approach was proposed in [5], however, rather than allocating all the time slots to periodic beacons, it dedicates some slots for event-driven messages. Additionally, it incorporates the use of relayers and introduces a relay selection policy tailored for platooning systems. Leveraging Slotted beaconing, Segata et al. [6] designed Jerk Beaconing in which the beaconing interval is dynamically adjusted based on Jerk, the rate of change of acceleration. To enhance reliability, Jerk Beaconing also uses acknowledgements and retransmissions. In [7], Balador et al. considered a different approach which is token-based where token management is centralised within a designated platoon member situated mid-platoon. This member generates the initial token, regenerates a new token upon loss and manages the token passing operation. Since their approach is based on the age of information, each platoon member maintains an updated list of all other platoon members and their last received beacons. The token is always passed to the vehicle with the oldest received data.

Our approach reduces the CBR while enhancing reactivity in critical scenarios such as emergency braking using intent beacons. It also prioritises simplicity by avoiding the use of retransmissions and without increasing the size of beacons. Integrating the proposed functionalities with the existing schemes could enhance their performance but it does not imply that they will become less complex.

Algorithm 1. PlatoonB
___
Require: $deceleration_{max}$, $actuationLag$, BI_{min} (minimum beaconing interval), BI_{max} (maximum beaconing interval), CBR_{limit} (desired CBR)
Ensure: Beaconing Interval BI in $[BI_{min} - \epsilon, min(BI_{max}, (TTC - actuationLag)) + \epsilon]$
1: $TTC \leftarrow getTTC()$ {approximate time to crash without actuation lag, from eq. 2}
2: $acceleration_{current} \leftarrow getCurrentAcceleration()$
3: $CBR_{current} \leftarrow getCurrentCBR()$
4: $BI_{max} \leftarrow max(BI_{min}, min(BI_{max}, TTC - actuationLag))$
5: $BI_{new} \leftarrow BI_{max} \cdot \left(1 - \left|\frac{acceleration_{current}}{deceleration_{max}}\right|\right)$
6: $b \leftarrow max\left(0, \left(\frac{CBR_{current}}{CBR_{limit}} - 1\right)\right)$
7: $BI_{new} \leftarrow BI_{new} + b \cdot BI_{new}$
8: $BI \leftarrow max(BI_{min}, min(BI_{max}, BI_{new}))$
9: $BI \leftarrow U(BI - \epsilon, BI + \epsilon)$ {offset. e.g., 0.001s}
10: $schedule(t_{current} + 0.05s, updateBeaconingInterval)$
___

3 Proposed Beaconing Schemes

3.1 System Model

We consider a number of platoons cruising on a one-directional, multi-lane highway. Each platoon consists of a leader and a number of followers. The leader is driven by a traditional adaptive cruise controller (ACC) which uses sensor readings (e.g., LIDAR) to automatically adjust the speed to maintain a safe distance from the vehicle or hazard ahead. Therefore, the leader does not require input from its followers. In contrast, the following vehicles are equipped with a Cooperative Adaptive Cruise Controller, specifically employing the PATH CACC controller which requires vital information, such as speed and acceleration, from the platoon leader and the immediate vehicle ahead. PATH CACC maintains a constant inter-vehicle distance d_d by outputting the desired acceleration u_i to be fed to the vehicle actuator, using the following control law (Eq. 1) [8]:

$$u_i = \alpha_1 u_{i-1} + \alpha_2 u_0 + \alpha_3(-d_{radar} + d_d) + \alpha_4(\dot{x}_i - \dot{x}_0) + \alpha_5(\dot{x}_i - \dot{x}_{i-1}), \quad (1)$$

where i is the vehicle index and $\dot{x}_i$ is its speed. The α_i constants are configuration parameters. d_{radar} is the radar-measured distance to front vehicle. $\dot{u}_{i-1}$ and $\dot{x}_{i-1}$ are the acceleration and speed of the front vehicle, respectively. u_0 and $\dot{x}_0$ are the acceleration and speed of the platoon leader, respectively. The acceleration and speed are obtained using the vehicle-to-vehicle wireless network.

All the vehicles are equipped with 802.11p network interface cards and they transmit their messages at 100mW of transmit power, propagating according to the free space model. The vehicles broadcast beacons according to the beaconing scheme they are running. Once a vehicle receives a beacon from the leader or from the vehicle ahead, it extracts the required information and feeds that to its CACC controller so that it outputs the required acceleration or deceleration to be in turn fed to a lower actuation controller which controls the vehicle's accelerator and brake.

Table 1. Algorithm 1 parameters

Parameter	Description
BI_{min}	Minimum beaconing interval (Typically 0.1 s or less).
BI_{max}	Maximum beaconing interval. It impacts awareness. (Typically 1 s or less).
$deceleration_{max}$	Maximum deceleration capability of a vehicle. It depends on vehicle type (e.g., -6 m/s^2).
$acceleration_{current}$	Current acceleration or deceleration of a vehicle.
$actuationLag$	Engine and braking lag. It varies. (e.g., 0.5 s).
CBR_{limit}	Desired threshold for the CBR to limit the channel usage of a specific application (e.g., 50%).
$CBR_{current}$	Current CBR measured over a period, such as 1 s.
ϵ	Offset for the randomisation of beacon generation at the application layer (e.g., 0.001 s).
TTC	Time to Crash obtained using Eq. 2.

3.2 PlatoonB

We propose Platoon Beaconing, PlatoonB, a distributed scheme that runs on each platoon member to reduce the communication channel load without negatively impacting the efficiency of the platooning application. The scheme is specified in Algorithm 1 which expects the following inputs: $deceleration_{max}$, the maximum deceleration capability of a platoon member; $actuationLag$, the delay taken by vehicle actuators (e.g., engine) to receive and execute the controller commands; BI_{min}, the desired minimum beaconing interval; BI_{max}, the desired maximum beaconing interval; CBR_{limit}, the desired CBR (Table 1). The scheme starts by calculating TTC which is an approximate time to crash to the vehicle ahead assuming the ego vehicle takes no action while the preceding vehicle brakes with maximum deceleration. This safety metric is used as an upper bound on the defined maximum beaconing interval BI_{max} after accounting for the actuation lag to ensure that the beaconing interval decreases as the inter-vehicle distance decreases (line 4 in Algorithm 1). Note that this metric is different from the headway time and TTC which are explained in [12]. TTC is estimated based on a basic equation of motion (Eq. 2) where $speed$ and $speed_{front}$ are the current speeds of the ego vehicle and the one in front, respectively. $Deceleration_{max}$ represents the maximum deceleration capability of the front vehicle, and d denotes the distance between the two vehicles. An example of how TTC reacts to different inter-vehicle distances is as follows: At a speed of 100 km/h and a maximum deceleration of -6 m/s^2, TTC (before subtracting the actuation lag) will be approximately 0.6 s, 1.3 s, and 1.8 s for inter-vehicle distances d of 1 m, 5 m, and 10 m, respectively.

$$speed \cdot TTC = d + speed_{front} \cdot TTC + \frac{1}{2} \cdot deceleration_{max} \cdot TTC^2 \quad (2)$$

Then, the scheme obtains the current acceleration of the ego vehicle and measures $CBR_{current}$, the current CBR. In line 5 of the algorithm, the beaconing interval BI_{new} will remain at its maximum value unless the vehicle speed changes ($acceleration_{current} \neq 0$). BI_{new} will decrease based on how close the current acceleration or deceleration is to the maximum value. The $|deceleration_{max}|$ is used as the maximum for both as it is reasonably assumed to be always greater than $acceleration_{max}$. The beaconing interval is reduced when a vehicle's speed changes which means more frequent and timely updates. This ensures that with

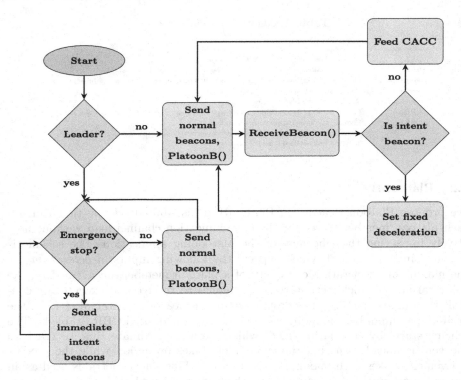

Fig. 1. PlatoonBE flowchart.

greater speed changes the vehicles send more beacons to maintain the safety of the platoon.

The beaconing interval is also impacted by the current communication channel state when it crosses a defined limit, CBR_{limit}. Below this limit, the beaconing interval is only affected by the change in acceleration or deceleration. That is, when the $CBR_{current} \leq CBR_{limit}$, b will be zero, otherwise b will increase BI_{new} to reduce the channel load. The final beaconing interval is uniformly selected within a specified offset range to further minimise simultaneous beaconing attempts at the application layer. The scheme schedules the beaconing interval for update every time t (e.g., 50ms) which has to be at most BI_{min} to allow the platooning controller to respond to changes promptly and safely.

3.3 PlatoonBE

As we will see in Sect. 4.2, all the evaluated schemes perform poorly during an emergency breaking scenario. This is because the beaconing approach presents a snapshot of speed and position that requires a subsequent beacon to be updated. This snapshot quickly becomes out of date during periods of rapid deceleration. This scenario poses a challenge for heavy-duty vehicles as they can decelerate at high rates ($6\,m/s^2$ or more) [13]. It is therefore logical to prioritise enhancements targeting this specific scenario to improve overall safety and performance. Therefore, we introduce PlatoonBE (PlatoonB Enhanced) where special beacons are sent immediately when the leader knows that it is going to perform an emergency braking at a specific deceleration rate. These beacons are not based on the beaconing interval. This is based on a reasonable assumption that the leader would know that it is going to perform an emergency braking due to a hazard or stopped traffic. The additional beacons include a special field identifying them as (intent beacons) as well as a field for the deceleration value. Once received by a follower, it will immediately perform the deceleration regardless of the output of its CACC controller. Note that the intents and the corresponding actions upon receiving these intents are predefined and pre-agreed between the leader and all its followers. That is, upon receiving an emergency braking intent, all the following vehicles will brake at the same deceleration rate as the leader.

Figure 1 illustrates the operation of PlatoonBE. If the ego vehicle is a leader, it follows the PlatoonB scheme for beaconing. If a leader determines that it needs to brake, it immediately sends out intent beacons. To increase the chances of successful intent delivery, redundant intent beacons can be sent. In this work, three beacons were chosen as a compromise between reliability and network overhead. If the ego vehicle is a follower, it will also continue beaconing as usual according to PlatoonB. However, upon receiving a beacon, a follower will check its type. If the beacon is a normal one from the platoon leader or the vehicle in front, the necessary information will be extracted and fed to the vehicle's CACC controller. On the other hand, if the beacon is an intent beacon, the encapsulated deceleration value will be extracted and executed immediately. This leads to a faster reaction time and therefore better inter-vehicle distances, as the results show in Sect. 4.

4 Evaluation

4.1 Metrics

To evaluate the impact of our approach on the V2V network, we use CBR, channel busy ratio which is the average percentage of time a channel is sensed busy by the physical layer. Lowering CBR is an important objective to reduce delay and packet loss as the channel is shared among the vehicles in the communication range. We also count the number of physical layer collisions per second to assess the overall performance of the network when our scheme is used as compared to the other schemes.

To assess the impact on platoon safety, we use the minimum inter-vehicle distance as our main metric. The closer the inter-vehicle distance to a predefined value (5 m in our scenarios), the safer the platoon is considered. In addition, we show the beaconing interval over simulation time for each scheme to understand its behaviour.

4.2 Experiments and Results

We implemented our schemes and conducted extensive simulations using PLEXE [2], a platooning framework for OMNeT++ and SUMO, to evaluate their performance compared to the other schemes. Regarding the DynB scheme, we observed that it was not suitable for platooning as it resulted in vehicle crashes in most of the simulation runs. This is because DynB uses the number of neighbours to rapidly increase the beaconing interval in response to even minor rises in CBR beyond the preferred threshold. Therefore, we instead used two variations of DynB that are more suitable for platooning: DynB1 and DynB2. DynB1 uses the same formula as DynB but additionally takes the logarithm base 2 of the neighbour count to reduce its impact on the beaconing interval. DynB2 uses only the parameter r as the beaconing interval, capped within the range $[BI_{min}, BI_{max}]$. Table 2 lists the simulation parameters.

In all the following experiments, the intra-platoon separation is 5 m while the inter-platoon separation is approximately 33.33 m. Distances of five meters or less between platoon members have been shown to result in much less energy consumption and CO2 emissions [9,10]. The inter-platoon separation is maintained by the ACC controllers on the leaders and is calculated as the product of headway and speed: $1.2s \cdot 27.78$ m/s. Each experiment includes two scenarios: low congestion (120 vehicles), and high congestion (480 vehicles). In the low congestion case we use 8 platoons while in the high congestion case we expand the number of platoons to 32, all of size 15 vehicles which is the size recommended by the SARTRE project [11]. The initial speed used in all experiments is 27.78 m/s, (100 km/h). In the Static and Slotted schemes, the beaconing interval is 0.1 s while in the other schemes, the beaconing interval is dynamic.

Experiment I (Normal). In this experiment, we compare the schemes under a normal or near-ideal platoon operation with very minor disturbances. Specifically, at 10.0 s of simulation time, the leaders slightly change their speed from 100 km/h to 100.5 km/h which results in an acceleration rate of 0.14 m/s^2 for a duration of 2 simulation seconds. The purpose of this experiment is to observe that the different approaches perform as expected.

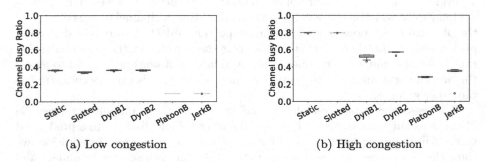

Fig. 2. Channel busy ratio [Experiment I].

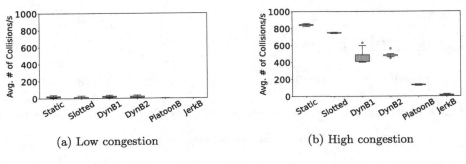

Fig. 3. Average number of collisions per second [Experiment I].

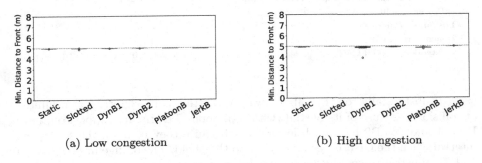

Fig. 4. Minimum distance to front vehicle [Experiment I].

Low Congestion (120 Vehicles). In this scenario, eight platoons were used. As shown in Fig. 2a, this led to around 35% of CBR for Static, DynB1 and DynB2. Here, DynB1 and DynB2 behaved exactly the same as Static (broadcasting at a fixed beaconing interval) since the CBR remained below the desired threshold, which we set to 50% throughout the simulation time. This is clearly demonstrated by the behaviour of the beaconing interval over time under DynB1 and DynB2, as depicted by the blue lines in a segment of the simulation presented in Fig. 5a and 5b. Slotted had nearly a similar figure since it also broadcasts 10 beacons per second while PlatoonB and JerkB showed a significantly less CBR, at about 9%. This is because both schemes always send beacons at the maximum interval (400ms) as long as the changes in speed are slight as in this experiment, as shown in Fig. 5c and 5d (note that the blue line is covered by the orange line). The number of packet collisions per second was low for all the schemes as the shared channel was not heavily congested in this scenario (Fig. 3a). With regard to the minimum inter-vehicle distance (Fig. 4a), as expected when disturbances are nearly absent, all the schemes achieved an almost perfect inter-vehicle distance.

High Congestion (480 Vehicles). With the increase in contention for the shared communication channel, the CBR jumped to about 80% for both the Static and Slotted schemes (Fig. 2b), and as expected, the number of collisions per second

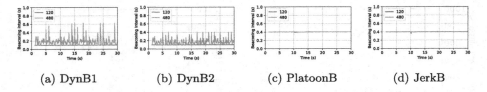

(a) DynB1   (b) DynB2   (c) PlatoonB   (d) JerkB

Fig. 5. Beaconing interval behaviour over time. [Experiment I]. (Color figure online)

Table 2. Simulation parameters.

Parameter	Value	Parameter	Value
Phy & MAC	802.11p (6 Mbit/s)	**Vehicle count**	120, 480
txPower	100 (mW)	**Platoon size**	15
Loss model	Free space	$deceleration_{max}$	6 (m/s^2)
Beacon size	200B	$actuationLag$	0.5 (s)
Controller	PATH CACC	$BI_{min}, BI_{max}, CBR_{limit}, p, \Delta u_{max}$	0.1, 0.4 (s), 0.5, 1, 2 (m/s^2)

surged substantially too (Fig. 3b). However, Slotted had a slightly lower number of collisions because of its partial (within platoon) TDMA. In comparison, both DynB1 and DynB2 had much lower CBR because they increase the beaconing interval when the CBR exceeds the desired threshold. The orange lines in Fig. 5a and 5b illustrate how the beaconing interval responds to variations in CBR. When the CBR exceeds 50%, the beaconing interval increases which leads to less traffic generated by vehicles, which in turn results in a lower CBR. Consequently, the beaconing interval decreases again and this cycle continues. Unlike DynB2, DynB1 also considers the $\log_2$ of the number of neighbours and this causes its beaconing interval to increase slightly more. This leads to a lower overall CBR for DynB1. PlatoonB demonstrated a notably better CBR, achieving around 28%, followed by JerkB with minimal increase in the number of collisions per second for both in contrast to the other schemes, with JerkB achieving the lowest number of packet collisions as it works on top of Slotted. Again, this is because in this experiment, both PlatoonB and JerkB send at the maximum beaconing interval almost all the time (Fig. 5c and 5d). Concerning safety, almost similar to the low congestion scenario, all the schemes had figures close to the desired distance of 5 m. This shows that even with high congestion, when the disturbances are very minor, all the schemes are very safe.

Experiment II (Oscillation). In this experiment, the goal is to determine whether a scheme can provide timely updates and tackle network congestion which can lead to packet loss and delays. Such issues can lead to unstable platooning. In an unstable platoon, vehicles are likely to exhibit high relative speeds leading to vehicle crashes. Initially, all platoons cruise at a speed of 27.78 m/s, and at 5.0 s of simulation time, the leaders of the first platoons start speeding up and then slowing down every 5.0 s at ±2.78 m/s.

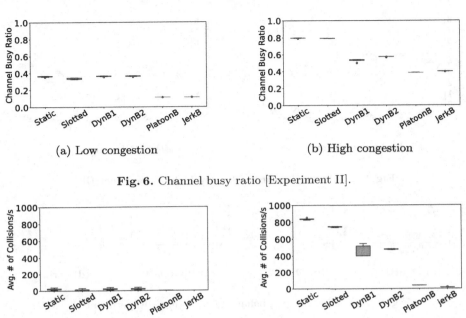

Fig. 6. Channel busy ratio [Experiment II].

Fig. 7. Average number of collisions per second [Experiment II].

Low Congestion (120 Vehicles). As shown in Fig. 6a, the CBR results for all schemes except PlatoonB and JerkB are similar to those observed in Experiment I. This arises because the other schemes do not react to changes in vehicle dynamics (variations in speed across different experiments). Therefore, their beaconing intervals remain unchanged from those in Experiment I (Fig. 9a and 9b). In contrast, PlatoonB and JerkB adapt to speed changes by dynamically reducing the beaconing interval and this led to a higher number of beacons sent and a slightly increased CBR. The collision figures remained unchanged across all schemes because the congestion was negligible as mentioned before, and besides PlatoonB and JerkB, the schemes did not alter their transmission rates.

With regard to the minimum inter-vehicle distance (Fig. 8a), all the schemes overall exhibited distances near the desired 5-meter. However, all other schemes experienced instances of distances falling below 1 m, with Slotted, DynB1 and DynB2 even resulting in at least one crash, none of which were observed with PlatoonB and JerkB, suggesting that they may be considered the safest schemes in this scenario.

High Congestion (480 Vehicles). As in Experiment I, the increase in contention for the shared communication channel led to a significant rise in CBR and the number of collisions per second. This resulted in figures almost identical to those in Experiment I for all schemes except PlatoonB (Figs. 6b and 7b) whose CBR

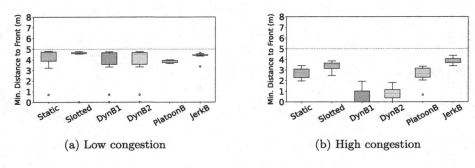

Fig. 8. Minimum distance to front vehicle [Experiment II].

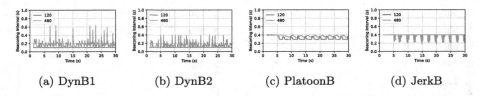

Fig. 9. Beaconing interval behaviour over time. [Experiment II].

increased to about 38% compared to about 28% in Experiment I, yet lower than that of JerkB. However, as shown in Figs. 9c and 9d, JerkB was more reactive. This is reflected in the inter-vehicle distance where JerkB achieved the best results followed by Slotted, PlatoonB and Static. Both DynB1 and DynB2 resulted in unsafe inter-vehicle distances with DynB1 being the poorest scheme in this scenario. This confirms that varying the beaconing interval dynamically based on the number of neighbours or CBR without considering the vehicle dynamics and the platooning application requirements is not a safe approach.

Experiment III (Stopping). This experiment tests the schemes in a critical situation, specifically emergency braking. At 5.0 s of simulation time, all platoon leaders brake at $-6m/s^2$ to a complete stop. Here, PlatoonBE is included as well.

Low Congestion (120 Vehicles) The figures for CBR and the number of collisions per second remained consistent with those in the previous experiments (Figs. 10a and 11a). In JerkB, more beacons were sent during sudden changes in speed (at 5.0 s and 10.0 s), as shown in Fig. 13d. In PlatoonB and PlatoonBE, more beacons were sent during the entire braking period, with PlatoonBE sending intent beacons at 5.0 s (Fig. 13c). The distances for all schemes were overall shorter compared to the previous experiments. Despite this, overall, the distances remained within acceptable limits. Notably, only JerkB, PlatoonBE and PlatoonB had no instances of vehicle crashes which shows that they are the safest in this scenario.

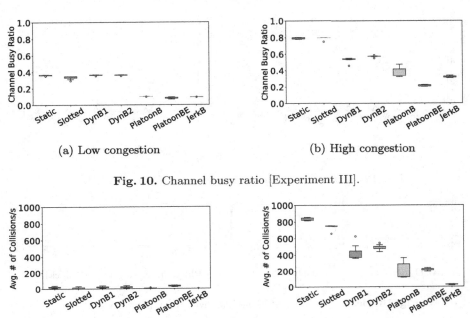

Fig. 10. Channel busy ratio [Experiment III].

Fig. 11. Average number of collisions per second [Experiment III].

High Congestion (480 Vehicles). As previously explained, Static and Slotted use fixed beaconing intervals, while DynB1 and DynB2 adjust their beaconing intervals independently of vehicle speed. Therefore, the CBR and the number of collisions per second for all these schemes remained nearly identical to the previous experiments (Fig. 10b and 11b). For PlatoonB, the CBR was overall lower but with greater variance compared to earlier results. This is because, for most of the simulation, the scheme operated at the maximum beaconing interval due to stable speeds (resulting in a low CBR), and only during the short braking period, the beaconing interval was at its minimum (resulting in a temporary high CBR, as shown in Fig. 13c). This fluctuation is also the reason for the increased and more variable number of collisions. PlatoonBE's results were similar to those of PlatoonB. However, due to higher synchronisation in transmission timing among vehicles, PlatoonBE experienced a higher number of collisions. This synchronisation led to more frequent instances where the channel was perceived as idle simultaneously, resulting in a lower CBR.

For safety, JerkB consistently resulted in vehicle crashes showing that it is ineffective when sudden braking occurs by multiple platoon leaders in high vehicle density situations (Fig. 12b). PlatoonBE achieved unparalleled safe distances, and therefore it stands out as the best and safest scheme in this scenario. Besides Slotted and PlatoonB, the remaining schemes exhibited impractical and unsafe distances.

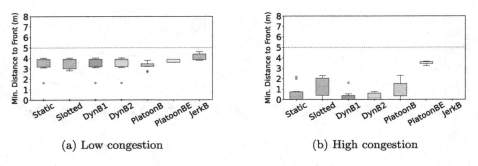

Fig. 12. Minimum distance to front vehicle [Experiment III].

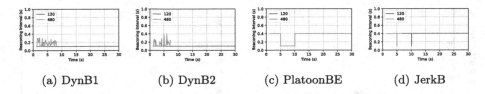

Fig. 13. Beaconing interval behaviour over time. [Experiment III].

5 Conclusion and Future Work

In this paper, we introduced PlatoonB and its enhanced version, PlatoonBE, a beaconing scheme that is light on the vehicle-to-vehicle communication channel which makes it suitable for platooning in congested scenarios, especially during critical situations. In summary, the results indicate that using a fixed interval for beaconing is impractical in heavy congestion scenarios unless mechanisms like TDMA are employed to mitigate packet collisions. Additionally, we observed that transmission rate control based solely on network indicators, such as the the number of neighbours or CBR, without considering application requirements is not effective. To be effective, it has to also consider vehicle dynamics and application requirements as done by JerkB and PlatoonB, and it performs even better when complemented with notifications as demonstrated in the enhanced version, PlatoonBE. Future work will take into account the variation in packet sizes across the different schemes and the related overhead because JerkB generates larger packets due to the inclusion of an acknowledgment map in each beacon. This was not considered in the current simulations and will be explored in the future. Future work will also focus on enhancing the reactivity of PlatoonBE in less critical scenarios and integrating various schemes to leverage their strengths. Further, the schemes will be evaluated when the underlying communication technology is C-V2X rather than 802.11p.

References

1. Balador, A., Bazzi, A., Hernandez-Jayo, U., Iglesia, I., Ahmadvand, H.: A survey on vehicular communication for cooperative truck platooning application. Veh. Commun. **35**, 100460 (2022)
2. Segata, M., Joerer, S., Bloessl, B., Sommer, C., Dressler, F., Cigno, R.: A platooning extension for Veins. In: 2014 IEEE Vehicular Networking Conference (VNC), pp. 53–60. IEEE, Paderborn (2014)
3. Sommer, C., Joerer, S., Segata, M., Tonguz, O., Cigno, R., Dressler, F.: How shadowing hurts vehicular communications and how dynamic beaconing can help. IEEE Trans. Mob. Comput. **14**(7), 1411–1421 (2015)
4. Segata, M., et al.: Toward communication strategies for platooning: simulative and experimental evaluation. IEEE Trans. Veh. Technol. **64**(12), 5411–5423 (2015)
5. Hoang, L., Uhlemann, E., Jonsson, M.: An efficient message dissemination technique in platooning applications. IEEE Commun. Lett. **19**(6), 1017–1020 (2015)
6. Segata, M., Dressler, F., Cigno, R.: Jerk Beaconing: a dynamic approach to platooning. In: 2015 IEEE Vehicular Networking Conference (VNC), pp. 135–142. IEEE, Kyoto (2015)
7. Balador, A., et al.: Supporting beacon and event-driven messages in vehicular platoons through token-based strategies. Sensors **18**(4), 955 (2018)
8. Segata, M., et al.: On platooning control using IEEE 802.11p in conjunction with visible light communications. In: Proceedings of the 2016 12th Annual Conference on Wireless On-demand Network Systems and Services (WONS), pp. 1–4. IEEE, Cortina d'Ampezzo (2016)
9. Tsugawa, S.: Results and issues of an automated truck platoon within the energy ITS project. In: IEEE Intelligent Vehicles Symposium Proceedings, pp. 642-647. IEEE, Dearborn (2014)
10. Murthy, D.K., Masrur, A.: Braking in close following platoons: the law of the weakest. In: Euromicro Conference on Digital System Design (DSD), pp. 613-620. IEEE, Limassol (2016)
11. Robinson, T., Chan, E., Coelingh, E.: Operating platoons on public motorways: an introduction to the SARTRE platooning programme. In: 17th World Congress on Intelligent Transport Systems, pp. 12. (2010)
12. Vogel, K.: A comparison of headway and time to collision as safety indicators. Accid. Anal. Prev. **35**(3), 427–33 (2003)
13. Zheng, R., et al.: Study on emergency-avoidance braking for the automatic platooning of trucks. IEEE Trans. Intell. Transp. Syst. **15**(4), 1748–1757 (2014)
14. Van Eenennaam, M., Wolterink, W.K., Karagiannis, G., Heijenk, G.: Exploring the solution space of beaconing in VANETs. In: IEEE Vehicular Networking Conference (VNC), pp. 1–8. IEEE, Tokyo (2009)
15. Bansal, G., Cheng, B., Rostami, A., Sjoberg, K., Kenney, J.B., Gruteser, M.: Comparing LIMERIC and DCC approaches for VANET channel congestion control. In: IEEE 6th International Symposium on Wireless Vehicular Communications (WiVeC 2014), pp. 1–7. IEEE, Vancouver (2014)
16. Bansal, G., Kenney, J.B., Rohrs, C.E.: LIMERIC: a linear adaptive message rate algorithm for DSRC congestion control. IEEE Trans. Veh. Technol. **62**(9), 4182–4197 (2013)

17. Dey, K., et al.: A review of communication, driver characteristics, and controls aspects of cooperative adaptive cruise control (CACC). IEEE Trans. Intell. Transp. Syst. **17**(2), 491–509 (2016)
18. Thandavarayan, G., Sepulcre, M., Gozalvez, J.: Cooperative perception for connected and automated vehicles: evaluation and impact of congestion control. IEEE Access **8**, 197665–197683 (2020)

Implementations Based Evaluation of No-Wait Approach for Resolving Conflicts in Databases

Yingming Wang[1], Paul Ezhilchelvan[1](✉), Jack Waudby[2], and Jim Webber[2]

[1] School of Computing, Newcastle University, Newcastle upon Tyne NE4 5TG, UK
{Y.Wang303,paul.ezhilchelvan}@ncl.ac.uk

[2] Neo4j UK, Union House, 182-194 Union Street, London SE1 0LH, UK
{Jack.Waudby,Jim.Webber}@neo4j.com

Abstract. In this paper, we describe No-Wait concurrency control mechanisms to address conflict resolution and then comprehensively evaluate their performance under Read-Committed and Serializability isolation levels using an in-memory database system in various configurations and contention scenarios. Key performance metrics are percentage of transaction aborts and average latency for those who do not abort. Our evaluations affirm that the No-Wait approach indeed offers a cost-effective, practical alternative to traditional conflict resolution mechanisms.

Keywords: Databases · Transactions · Concurrency control · Deadlock Avoidance · No-Wait · Implementation · Performance Evaluation

1 Introduction

In traditional Database Management Systems (DBMS), managing concurrent transactions often leads to blocking, where a transaction is made to wait for another ongoing transaction when it encounters an access *conflict* over the same data, causing performance issues. The most challenging issue is a *deadlock* - a situation where transactions wait for each other, halting progress altogether. Deadlock treatment strategy is either *preemptive*, avoiding deadlocks before they happen, or *reactive*, resolving them after they occur.

There are three types of access conflicts in a DBMS: after an on-going transaction has written a data item, say, X, if another one wishes to write or read X, then a *write-write* or *write-read* conflict is said to occur, respectively; similarly, a *read-write* conflict arises when a later transaction seeks to write X after an earlier one has read X. An in-depth literature review on the topic of conflict management in database systems can be found in our previous work [1].

The canonical work on deadlock management, *Serialization Graph Testing* (SGT) [2], dynamically constructs a graph wherein nodes represent ongoing transactions and edges the conflicts among these transactions. Cycles in this graph indicate deadlocks. By aborting only those transactions that need to be aborted, deadlocks are resolved efficiently. Thus, SGT keeps the number of transaction aborts to the bare minimum. However, SGT incurs a significant overhead due to the need to maintain a graph and is not well-suited to distributed DBMS. Since distributed transactions are a focus in our on-going work, we will use SGT as a baseline for comparing the performance of the protocols presented here.

In this paper, the term *No-Wait* will refer to a non-blocking concurrency control strategy, designed to prevent transactions from waiting on conflicts and eliminate thereby any risk of deadlocks as well as to yield high-performance (see Li et al. [3]). Consider an object that is already accessed by an ongoing transaction, say, *T1*, and needs to be accessed also by another transaction, say *T2*, giving rise to a conflict. In No-Wait, *T2* never waits for the lock to be released by *T1* but either *aborts* itself or *wounds T1* by forcing *T1* to abort; i.e., every conflict encountered is *pessimistically* assumed to lead to a deadlock if one transaction is made to wait for another to complete, and is dealt with by ensuring that no transaction waits for locked data to be released. There are two options for *T2* to abort itself or wound *T1*: *instantaneous* or *delayed*.

In the *instantaneous* option, conflicts are addressed immediately upon being encountered. Upon detecting a conflict, an incoming transaction immediately *wounds* the preceding transaction or *aborts* itself. It has been analytically evaluated and simulated in our earlier work [1]. In the *delayed* option, a transaction encountering a conflict delays conflict resolution and proceeds with its operation without waiting, thereby leaving the conflict unresolved *for now*. After it completes all its operations, it enters a *validation* phase and resolves each conflict encountered earlier (by wounding or aborting), *if* the conflict still exists during validation. Note that a conflict disappears if a preceding transaction completes before a later transaction validates. Thus, delayed No-Wait optimistically expects conflicts encountered earlier to naturally disappear with passing of time.

Assessing the performance impact of both instantaneous and delayed versions of No-Wait using benchmark-compliant implementations on a multi-core machine is the main contribution of this paper while [1] analytically evaluated and simulated only instantaneous versions. The fraction of transactions that abort and completion times of those that do not abort are the metrics measured and compared. Further, our performance study considers the two most widely used isolation levels [2]: *Read-Committed* and *Serializability*. Under Read-Committed, we consider both instantaneous and delayed options. Under Serializability, however, practical considerations force us to employ only the instantaneous option and to introduce several *hybrids* to address specific types of conflicts.

2 No-Wait Under Read-Committed

At the Read-Committed isolation level, a transaction can only read data that has already been committed by other transactions, effectively preventing it from reading *dirty* value that has been written by an on-going (and hence uncommitted) transaction. At this isolation level, only write-write conflicts can occur.

1. Instantaneous Scheme
    - *Instantaneous Abort* (IA): The incoming transaction **aborts** itself, requiring a rollback, i.e., undoing any and all write operations it may have carried out earlier. The preceding transaction is left undisturbed.
    - *Instantaneous Wound* (IW): The incoming transaction **wounds** the preceding transaction by marking it as "aborted" and continues to write. In scenarios such as social media or short video applications [3], the incoming transaction may possess fresher information, making it advantageous to allow the later transaction to proceed. Compared to the Instantaneous Abort, this method requires more computational resources, as each transaction needs to continually check if it has been wounded.
2. Delayed Scheme
    - *Delayed Abort* (DA): The incoming transaction **delays aborting** itself on encountering a write-write conflict by proceeding with its own operations and additionally recording all active preceding transactions involved in that conflict. In the validation phase, it passes its validation *if and only if* all preceding transactions recorded earlier are complete; i.e., it aborts itself even if one of the recorded predecessors is found to be still active during its validation. DA, like IA, incurs a low overhead and, unlike IA, seeks to minimise the chances of an abort outcome.
    - *Delayed Wound* (DW): The incoming transaction records its predecessors as in DA. At the validation phase, it wounds all the recorded predecessors that are found to be still active by marking them as "aborted", ensuring thereby that active predecessors do not overwrite its written value later. On the other hand, if a predecessor is found to be aborted or committed during validation, nothing is done. Thus a transaction in DW seeks to minimise the number of other transactions it wounds and its own chances of being wounded.

Read-Committed (RC) permits the *lost update* issue, which occurs when two or more transactions read the same data item and then make independent updates based on the value they read. The following two cases exemplify this issue when RC is enforced, where RJ, WJ and CJ indicate read, write and commit by transaction TJ, respectively.

Case 1: R1[X=10] R2[X=10] W2[X=X+10] W1[X=X+20] C2 C1
Case 2: R1[X=10] R2[X=10] W2[X=X+10] C2 W1[X=X+20] C1

In both cases, transactions $T1$ and $T2$ read the same committed value of X, which is 10. $T2$ writes X as 20 followed by $T1$ writing X to 30. In Case 2, $T2$ writes and commits which is then followed by $T1$ writing, leaving X at 30. In both scenarios, the update X= X+10 by $T2$ is *lost* due to the subsequent write by $T1$. This lost update phenomenon results in a final data state that does not reflect the update of $T2$, but is still permissible under RC isolation level.

Since C2 occurring before C1 implies that $T2$ completes before $T1$, avoiding lost update must leave the final value of X as 40 in both cases. We illustrate below that neither our schemes nor SGT can guarantee avoidance of lost update:

1. IA/IW: These mechanisms resolve write-write conflicts instantaneously when multiple transactions seek to write the same data item. Thus, they avoid lost update in Case 1. However, in Case 2, after *T2* commits, there is no write-write conflict on X for *T1* to detect. Therefore, the instantaneous schemes allow lost update in Case 2.
2. DA/DW: They resolve write-write conflicts during the validation phase by either aborting itself or wounding the preceding transaction if any preceding transaction is still active. In Case 1, lost update is prevented, if *T1* validates before *T2* commits; otherwise, not. Neither DA nor DW can prevent lost update in Case 2. Thus, lost update can also occur in delayed schemes.
3. SGT: SGT can allow both cases of lost update in a manner similar to the delayed schemes.

3 No-Wait Under Serializability

Under serializability, conflicts of all three types must be handled to ensure that the concurrent execution of transactions remain equivalent to some serial execution [2]. While instantaneous schemes are readily applicable for conflicts of all types, delayed schemes are not so, because they can introduce multiple versions of data in the interval that elapses between write operations and the subsequent (intentionally delayed) no-wait conflict resolution during validation. The rationale for selecting appropriate delayed schemes is provided below, after presenting the instantaneous schemes and *hybrid* solutions that these selections led to:

1. Instantaneous Scheme
   - *Instantaneous Abort* (IA): The incoming transaction aborts itself instantaneously, when it encounters a conflict, irrespective of the type of conflict encountered: write-write, write-read, or read-write.
   - *Instantaneous Wound* (IW): The incoming transaction, on encountering a conflict of any type, wounds the preceding transaction instantaneously by changing its state to "aborted".
2. Hybrid Schemes
   - *Hybrid 1* (H1): It uses Instantaneous Abort for resolving write-write conflicts and delayed schemes to handle both the read-related conflicts: Delayed Wound for read-write conflicts and Delayed Abort for write-read conflicts. During the validation phase, all read related conflicts are considered. A transaction wounds its predecessor in a read-write conflict only if the read predecessor is still active; otherwise no wounding is done. Further, a transaction being validated aborts itself if its predecessor in a write-read conflict is found not to have already committed, i.e., if the write predecessor either has already aborted or is still running.
   - *Hybrid 2* (H2): The only difference between Hybrid 1 and Hybrid 2 is that Hybrid 2 uses Instantaneous Wound to resolve write-write conflict.

Table 1. Hybrid conflict-resolution approaches to Serializability

Conflict type	H1	H2	Notes
write-write	IA	IW	*instantaneously:* abort itself (H1) or wound predecessor (H2)
read-write	DW	DW	*during validation:* wound every active read predecessor
write-read	DA	DA	*in validation:* abort itself if write predecessor not committed

Table 1 outlines two hybrid solutions for addressing specific types of conflicts, which is then followed by the rationale for employing a particular method to resolve each type of conflict.

Only instantaneous schemes are employed for **write-write conflict resolution in hybrids**. The reason for not considering delayed schemes is that they permit simultaneous existence of multiple uncommitted writes on a given data item. At the Serializability isolation level, when several transactions modify a given data element, abort of one can cause all those transactions which later wrote that item, to abort as well. These *cascading aborts* increase system load and are not useful for both the user and the system [4]. Managing these cascading aborts requires substantial computational resources and memory. The DBMS must track the temporal dependencies among transactions to determine which ones must be aborted following a given transaction's abort. This tracking process, coupled with the execution of multiple transaction aborts and their associated rollbacks, imposes a heavy load on the system.

In addition, another consequence of cascading aborts is the temporary unavailability of the database segments affected by these aborts. This unavailability can severely degrade system throughput and negatively impact the user experience. Therefore, the delayed scheme, which increases the risk of cascading aborts, is excluded. The following examples provide a detailed explanation of how cascading aborts occur.

An example of using DW leading to cascading aborts is shown in Table 2 where tj denotes time step j. Data items X and Y are accessed by concurrent transactions, $T1$ to $T5$. This Table elucidates the sequence of operations executed by these concurrent transactions over a defined temporal span. Transaction operations are denoted as follows: $W(X)$ represents a read-then-write operation conducted on data X; $R(X)$ represents a read operation executed on data X; A

Table 2. Delayed Wound leads to cascading aborts.

Txn	t0	t1	t2	t3	t4	t5	t6	t7	t8	t9	t10	t11
T1	W(X)						A					
T2		W(X)					V	A				
T3			W(X)	W(Y)					A			
T4					R(Y)					A		
T5						W(Y)					A	

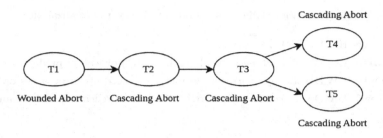

Fig. 1. Transactions' cascading aborts in the example

represents abort operation; V the validation phase that precedes commit operation which does not appear in Table 2 as no transaction commits (Fig. 1).

Initially, *T1* performs the first read-then-write operation on X at step $t0$. *T2* reads this newly written, uncommitted value of X in time step $t1$ and writes a second new value for X. In the next step $t2$, *T3* reads the value of X - the one written at step $t1$ - and writes a third new value for X. Subsequently, *T3* performs operations on data Y according to its transaction logic, writing a value to Y at time $t3$, which is then used by *T4* and *T5* at time steps $t4$ and $t5$ respectively.

Remark 1: When a transaction, say T, reads the value of an uncommitted data, say X, and then uses that value to compute a value to be written or X (or any other data item), the second write under Serializability must be aborted if the uncommitted value used in computation is not subsequently committed; otherwise, an equivalent serial execution cannot be derived. So, T must abort, irrespective of the conflict resolution scheme used under Serializability.

Remark 2: The issue of T in *Remark 1* aborting does not arise under RC, because T never reads an uncommitted value. However, as we have discussed earlier, reading only the committed value, instead of the latest uncommitted value, leads to lost update anomaly which cannot occur under Serializability.

Referring to Table 2, suppose that *T2* enters the validation phase first in time step $t6$ and wounds its active predecessor *T1* as per DW scheme. This leads to *T2* aborting itself as per *Remark* 1 which then triggers a cascade of

Fig. 2. read-write conflicts　　　　　**Fig. 3.** write-read conflict

all transactions aborting. A similar example showing cascading aborts can be constructed when DA is used.

Consequently, we exclude delayed schemes to resolve write-write conflicts, opting exclusively for instantaneous schemes. The choice between Instantaneous Abort and Instantaneous Wound depends on whether later transactions are assigned equal or higher priority as explained earlier.

DW is turns out to be the optimal strategy for handling **read-write conflicts** in hybrid schemes. Our decision to exclude instantaneous schemes is aimed to improving the throughput of *read-only* transactions without impeding update transactions. Illustrated by Fig. 2, $T4$'s write operation presents conflicts with three read-only transactions. The Instantaneous Wound would abort all preceding transactions, detrimentally affecting the throughput of read-only transactions. Additionally, the Instantaneous Abort, especially in the presence of frequently accessed "hot" data, forces update transactions into self-abort until the contested data item becomes idle, significantly compromising update transaction throughput and availability.

Thus, we determine the instantaneous scheme unsuitable. Additionally, the Delayed Abort, which necessitates self-abortion of update transactions upon encountering active preceding transactions during validation, is impractical due to the expensive rollback processes.

Delayed Wound thus emerges as the optimal strategy for handling read-write conflicts by deferring conflict resolution until its validation phase, particularly if the preceding transaction remains active. For instance, in Fig. 2, $T4$ will wound the still active transactions among $T1$, $T2$, and $T3$, in $T4$'s validation phase. This approach is predicated on the characteristic behavior of read-only transactions in a DBMS employing Delayed Wound: these transactions primarily require validation against dirty reads; they do not modify data. The Delayed Wound allows read-only transactions additional time to complete until its validation phase, thereby enhancing the overall throughput of read-only transactions. Moreover, it is highly probable that preceding transactions still active during the validation phase are update transactions, necessitating their abortion. On the other hand, the chance that an ongoing preceding transaction is a lengthy read-only transaction, is relatively small. Therefore, by adopting the Delayed Wound, DBMS affords read-only transactions the necessary leeway to complete without immediate interruption, ensuring a balanced throughput between read-only and update transactions.

For handling **write-read conflicts, DA** turns out be appropriate, as both wound strategies are unsuitable. It is illogical for read-only transactions, which do not modify any data, to have the capability to abort update transactions. Specifically, while the Instantaneous Wound approach may prevent dirty reads and has the potential to increase read-only transaction throughput, its benefits are overshadowed by the negative impact on update transaction throughput and the substantial rollback costs involved. On the other hand, the Delayed Wound introduces the risk of contaminating read values, consequently necessitating the abortion of both involved transactions. Furthermore, the remaining Instantaneous Abort impacts the throughput of read-only transactions, diverging from our objectives. Consequently, we conclude that Delayed Abort represents our preferred method for addressing write-read conflicts.

In scenarios of write-read conflicts and Delayed Abort, a transaction does a self-abortion during its validation phase if preceding transactions remain active or have already been aborted. Figure 3 exemplifies this with *T2* aborting itself should *T1*, the preceding transaction, be active or aborted. This strategy pessimistically avoids dirty reads without waiting for the completion of preceding transactions in the validation phase. Additionally, the Delayed Abort approach extends the temporal window for the completion of preceding transactions, thereby potentially enhancing the throughput of read-only transactions by providing them with a broader time frame to conclude successfully.

4 Experiments

This section presents our performance evaluation of the protocols discussed in Sect. 3. The experiments were conducted on a Microsoft Azure virtual machine using an E64s v5 instance type, which features 64 CPU cores and 512 GB of high-performance memory. To conduct our experiments, we created databases according to two standard benchmarks (see below) and the contents were filled in using appropriately chosen random values. The abort percentage and average response time of *committed* transactions are the metrics measured by varying the number of cores (and thus the maximum concurrent transactions) and the data access pattern by transactions.

Section 4.1 introduces the benchmarks used in our experiments: SmallBank [5] and Yahoo! Cloud Serving Benchmark (YCSB) [6]. The subsequent sections, 4.2 and 4.3, discuss the experimental results under Read-Committed and Serializability isolation levels, respectively.

4.1 Benchmarks and Workloads

The section describes the benchmarks deployed in our experiments.

SmallBank: The SmallBank benchmark [5] is a widely used microbenchmark for evaluating the performance in terms of transaction processing. It comprises three tables: *Account*, *Saving*, and *Checking*, each containing two columns, with the *account id* as the primary key. We have set the number of rows for each table as 1,000. SmallBank also specifies six transactions: *Balance*, *DepositChecking*, *TransactSaving*, *Amalgamate*, *WriteCheck*, and an additional *SendPayment*. Of these, only *Balance* is a read-only transaction, while the rest involve both read and write operations, following a read-then-write pattern. The number of operations for each transaction type is fixed between 3 and 8 (e.g., 3 and 8 for *Balance* and *Amalgamate* respectively). Finally, each transaction type is equally likely to occur.

YCSB: The YCSB (Yahoo! Cloud Serving Benchmark) [6] is employed to evaluate the performance of various cloud databases. YCSB includes a single table with 1,000 rows and 11 columns, where the first column is the primary key, and each of the other columns contains 100 bytes of random characters. In YCSB, there is only one type of transaction, which involves 10 accesses to randomly chosen data in a random order. Read and write operations occur with equal probability, 50%. Data access follows a *Zipf* distribution, with the frequency of access to randomly-chosen "hot" data controlled by a skew parameter, *theta*. When the *theta* is set to 0, there is no hot data and every data is accessed with uniform frequency.

4.2 Read-Committed

This section evaluates the protocols under Read-Committed isolation using the SmallBank and YCSB benchmarks. For the SmallBank benchmark, Figs. 4 and 5 illustrate the abort percentage and response time. Subsequently, Figs. 6 and 7 present the abort percentage and response time for the YCSB benchmark, focusing on scenarios with *theta* = 0 and a 50% update rate. The comparative analysis reveals that, in contrast to SmallBank's diverse transaction types with varying lengths, YCSB transactions are unique in type and consistently longer, involving more write operations. This results in a higher abort percentage and longer response times due to the increased likelihood of write-write conflicts.

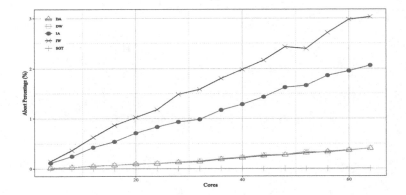

Fig. 4. Abort percentage under SmallBank

Figures 4 and 6 display the abort percentages under the two benchmarks. At the Read-Committed isolation level, various protocols exhibit distinct abort percentages due to their inherent conflict resolution strategies. The SGT protocol consistently demonstrates the lowest abort rate among these protocols, attributed to its comprehensive cycle detection mechanism. Even as the number of cores increases, SGT maintains an abort percentage close to 0%, with Figs. 4 and 6 showing an abort percentage of approximately 0.002% and 0.033%, respectively, at 64 cores.

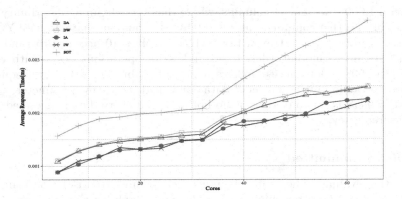

Fig. 5. Response time under SmallBank

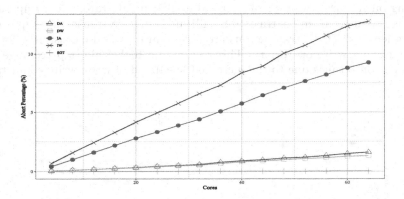

Fig. 6. Abort percentage under YCSB

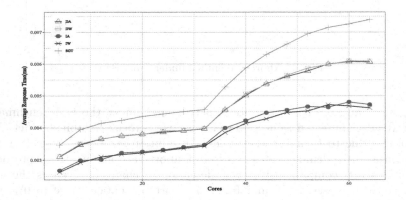

Fig. 7. Response time under YCSB

In contrast, the instantaneous schemes, including Instantaneous Abort (IA) and Instantaneous Wound (IW), exhibit higher abort percentages due to their immediate and pessimistic conflict resolution approach. These increases in abort percentage are more pronounced in the longer transactions of YCSB compared to those in SmallBank. Between these two protocols, the IW slightly increases the abort percentage compared to the IA. This difference primarily arises in scenarios where transactions can be wounded by others while attempting to wound or awaiting wound results. To prevent indefinite wounding attempts, transactions will self-abort after a limited number of retries.

The delayed scheme, including Delayed Abort (DA) and Delayed Wound (DW), positions itself between SGT and the instantaneous schemes by deferring conflict resolution to the validation phase. This approach reduces the abort percentage compared to the instantaneous schemes by allowing transactions to continue without immediate termination upon encountering a conflict.

Figures 5 and 7 depict the response times under the two benchmarks. SGT's cycle checks significantly elongate transaction response times. As the number of cores increases, SGT's response time rises more rapidly due to increased contention, which adds complexity to maintaining the dependency graph and checking for cycles. Additionally, the longer transactions in YCSB contribute to this complexity, leading to a steeper rise in response time compared to SmallBank.

The instantaneous schemes prioritize rapid conflict resolution, resulting in the shortest response times among the protocols. Despite their operational differences, IA and IW report similar response times. Theoretically, the slight edge in speed observed with the IA is due to it bypassing the overhead of modifying the status of preceding transactions, a step present in the IW process. However, in practice, the high-speed computation capabilities of modern computers and the limited number of wound attempts result in similar response times.

The delayed schemes find a middle ground in response times by deferring conflict resolution to the validation phase. This strategic delay allows them to reduce the overhead associated with instantaneous conflict resolution and the extensive cycle checks of SGT.

4.3 Serializability

Figures 8 and 9 display the abort percentages and response times for the protocols using the SmallBank benchmark. For the YCSB benchmark, Figs. 10 and 11 present these performance metrics under $theta = 0$ and a 50% update rate. Additionally, Figs. 12 and 13 show the performance metrics with varying skew values ($theta$), 32 cores, and a 50% update rate.

An analysis of abort percentages reveals a similar pattern across protocols compared to their performance under Read-Committed isolation. However, a notable distinction is that Serializability requires handling all types of conflicts to maintain data consistency, which significantly increases the likelihood of conflicts and consequently leads to a higher abort percentage compared to Read-Committed isolation.

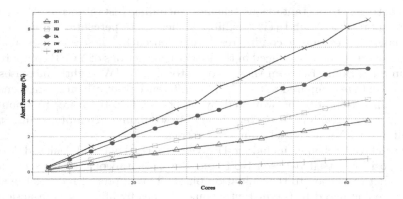

Fig. 8. Abort percentage under SmallBank

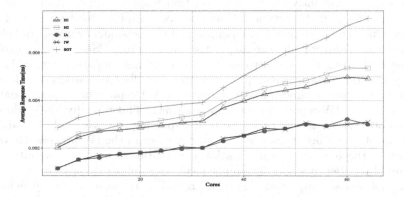

Fig. 9. Response time under SmallBank

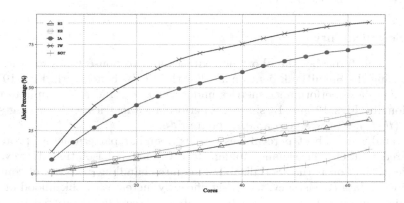

Fig. 10. Abort percentage under YCSB

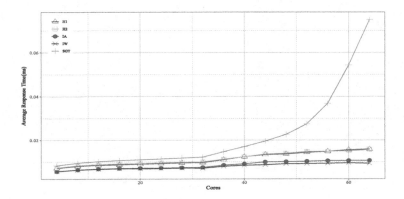

Fig. 11. Response time under YCSB

A significant rise in the abort percentage for SGT is observed in YCSB when the number of cores is very high. This increase is attributed to extremely high contention, which triggers numerous cycles, leading to transaction aborts and cascading aborts.

The instantaneous schemes register the highest abort percentages. At 64 cores, their abort percentages are approximately 7% for SmallBank and 75% for YCSB. Longer transactions require more data access and take more time to execute, increasing the probability of being aborted or wounded.

Positioned between these, Hybrid 1 (H1) and Hybrid 2 (H2) display similar abort percentages under both benchmarks, approximately 3% and 33% at 64 cores respectively. They postpone the resolution of read-related conflicts while directly addressing write-write conflicts, resulting in slow growth in abort percentages, even in high concurrency and long transaction scenarios. H2's abort percentage is marginally higher than H1's because wounding can lead to more cascading aborts and the potential for being wounded by another transaction during its wounding.

Regarding response times, the instantaneous schemes maintain the shortest durations, reflecting the efficiency of their conflict resolution methods. In contrast, SGT exhibits the longest response times due to its exhaustive cycle verification process. This effect is more pronounced under high contention, as evidenced in Figs. 9 and 11, where response times increase significantly after 32 cores. H1 and H2 occupy a middle ground, offering quicker response times than the rigorous SGT but slower than the instantaneous schemes. This is achieved by delaying conflict resolution and not waiting for the completion of preceding transactions. Even under high contention in YCSB, as shown in Fig. 11, H1 and H2 still demonstrate good response times with only slow growth.

We observed the performance metrics under varying skew values, *theta*, while keeping other parameters constant: 32 cores and a 50% update rate. Figure 12 illustrates the changes in abort percentages for these protocols. As expected, SGT maintains the lowest abort percentage until the skew value reaches 0.7. At

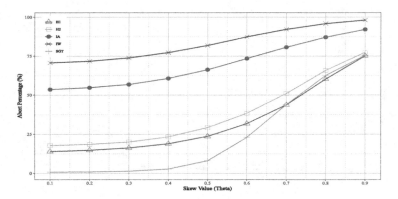

Fig. 12. Abort percentage with skewed value

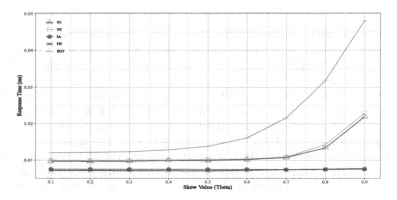

Fig. 13. Response time with varying skew values

higher skew values, transactions continually accessing highly contended data lead to persistent cycles, eventually causing transaction aborts and cascading aborts. This results in SGT's abort percentage approaching that of H1 beyond $theta = 0.7$. In contrast, IA and IW consistently exhibit high abort percentages, even at low skew levels. Positioned between these extremes, H1 and H2 show a steady increase in abort percentages. Their strategy of allowing preceding transactions time to complete results in the abort percentages under high contention that are similar to SGT's.

Figure 13 displays the response times as *theta* increases. SGT experiences a rapid rise in response time beyond $theta = 0.6$. This is due to skewed data access causing cycles to appear more frequently, increasing the complexity of maintaining the graph and detecting cycles, thereby leading to longer response times. In contrast, IA and IW show almost no change. In scenarios with a small data set and high contention, the variation in *theta* has little impact on the abort percentages of these two protocols. Positioned between them, H1 and H2 begin to show increased response times only after $theta = 0.7$, and their rate of increase is slightly smaller compared to SGT.

5 Conclusions

In this paper, we presented a detailed analysis of several transaction processing protocols and evaluated their performance under two benchmarks, SmallBank and YCSB. Delayed Abort and Hybrid 1 proved to be an excellent choice, under Read-Committed isolation and Serializability respectively, with the latter maintaining a good balance between abort percentage and response time, regardless of transaction length or data access skew. Our future work will use these schemes in a distributed setting.

References

1. Ezhilchelvan, P., Mitrani, I., Webber, J., Wang, Y.: Evaluating the performance impact of no-wait approach to resolving write conflicts in databases. In: Iacono, M., Scarpa, M., Barbierato, E., Serrano, S., Cerotti, D., Longo, F. (eds.) Computer Performance Engineering and Stochastic Modelling, pp. 171–185. Springer Nature Switzerland, Cham (2023). https://doi.org/10.1007/978-3-031-43185-2_12
2. Bernstein, P.A., Hadzilacos, V., Goodman, N.: Concurrency Control and Recovery in Database Systems. Addison-Wesley Longman Publishing Co., Inc, USA (1986)
3. Li, C., et al.: ByteGraph: a high-performance distributed graph database in bytedance. Proc. VLDB Endow. **15**(12), 3306–3318 (2022). https://doi.org/10.14778/3554821.3554824
4. Durner, D., Neumann, T.: No false negatives: accepting all useful schedules in a fast serializable many-core system. In: 2019 IEEE 35th International Conference on Data Engineering (ICDE), pp. 734–745 (2019)
5. Alomari, M., Cahill, M., Fekete, A., Rohm, U.: The cost of serializability on platforms that use snapshot isolation. In: 2008 IEEE 24th International Conference on Data Engineering, pp. 576–585 (2008)
6. Cooper, B.F., Silberstein, A., Tam, E.: Benchmarking cloud serving systems with YCSB. In: Proceedings of the 1st ACM Symposium on Cloud Computing, ser. SoCC 2010. New York, NY, USA: Association for Computing Machinery, 2010, pp. 143–154 (2010). https://doi.org/10.1145/1807128.1807152

Performance Evaluation of Smart Bin Systems Using Markovian Agents for Efficient Garbage Collection

Enrico Barbierato[1](✉), Alice Gatti[1], Marco Gribaudo[2], and Mauro Iacono[3]

[1] Dip. di Scienze Matematiche, Fisiche e Naturali, Università Cattolica del Sacro Cuore,
via della Garzetta 48, 25133 Brescia, Italy
{enrico.barbierato,alice.gatti}@unicatt.it

[2] Dip. di Elettronica, Informatica e Bioingegneria Politecnico di Milano,
via Ponzio 51, 20133 Milano, Italy
marco.gribaudo@polimi.it

[3] Dip. di Matematica e Fisica Università degli Studi della Campania "L. Vanvitelli",
viale Lincoln 5, 81100 Caserta, Italy
mauro.iacono@unicampania.it

Abstract. Smart bins, equipped with sensors and IoT technologies, play a crucial role in optimizing waste collection by providing real-time data on bin fill levels. This paper introduces a Markovian Agent Model to simulate and evaluate different garbage collection strategies in a smart bin system. By analyzing various alarm thresholds and routing policies, the study identifies optimal approaches for minimizing overflows and enhancing collection efficiency. The results demonstrate that a strategy combining responsive alarm handling with route resumption (Resume policy) and a higher alarm threshold improves system stability and operational effectiveness.

Keywords: markovian agents · smart bins · performance analysis

1 Introduction

In the contemporary landscape of smart city development, advanced technologies are leveraged to enhance efficiency, sustainability, and overall welfare. Therefore, multiple aspects of everyday life are characterized by innovative solutions enabling smartness. In the context of waste management systems, smart bins are intelligent waste containers that integrate IoT sensors, AI, and data analytics to optimize waste collection processes in sight of both environmental sustainability and efficiency. A smart bin is a technologically enhanced waste dumpster equipped with both sensors and technologies that enable advanced features and functionalities, improving the usual waste management processes. Each smart bin is typically assigned a unique identification (ID) to distinguish it from other bins within the system. Smart bins are integrated with Geographic Information Systems (GIS) for spatial data collection, analysis, and visualization purposes. Thus, real-time dynamical geographical data, traffic flows and conditions, and road networks are made available. In particular, each bin has precise geographical coordinates (latitude and longitude) associated. Merging the bins' real-time data with cities' roadmaps uncovers concentrations of waste generation, optimized collection routes, and enables a more efficient allocation of resources. Long-term planning and infrastructure development are also affected. Smart bins are equipped with sensors to measure and immediately record information about the waste level. Different types of sensors are available. Visual measurement is accomplished through ultrasonic sensors that measure the distance between the bin cover and its bottom via propagation of sound waves and consequently achieve a precise fullness measure via calculations on the expired time between the emission and the return of the waves. Weight measurement is instead accomplished by the usage of a weighing scale mounted on a double-bottom of the bin. Every bin records its present level of waste, signifying its capacity from being empty to completely overflowing. The sensors serve for the decision-making process. If a specific threshold is exceeded, the bin is added to the waste collection routing queue. A bin might be added to the routing schedule even if its waste threshold is not reached, as bins are also integrated with time-out mechanisms. A pre-determined time limit is set and, when triggered, the bin is automatically enrolled for collection. All smart bins are part of a network and are connected via wireless communication systems. Each bin is expected to send messages to the network, notifying its current state. Microcontrollers and microcomputer technologies enable integrated systems to perform local processing. Also, the real-time collected data can be stored and processed in a cloud-based system accessible via the bins' connection. External systems and applications can be integrated via the usage of Application Programming Interfaces (APIs). As each bin has a unique IP address in the network, APIs enable methodical and simultaneous queries of all the bins. Message Queuing Telemetry Transport (MQTT) allows the swift exchange of information through its lightweight design and publish-subscribe architecture, making it a suitable transmission method for the messages of the network of bins. Depending on the state of a bin, it can be dynamically inserted in route planning. However, the adaptive and real-time optimization of waste collection routes based on the continuously changing fill levels of smart bins requires dynamic routing. The collection schedule is dynamically adjusted for the routing of collection vehicles to bins that are marked as "to be emptied". Consequently, the unnecessary visits to bins with lower fill levels can be minimized, optimizing efficiency and reducing operational costs.

The contribution of this work lies in introducing a framework that allows garbage collectors to adapt dynamically between routine operations and emergency responses based on real-time data from smart bins. By evaluating different routing strategies and alarm thresholds, the study provides valuable insights into optimizing waste collection efficiency, minimizing overflow incidents, and enhancing system stability. The paper is organized as follows. Section 2 reviews the related work. Section 3 characterizes the considered case study and the corresponding model, which is simulated and discussed in Sect. 4. Finally, Sect. 5 concludes this work.

2 Related Work

A comprehensive review of classic and ML-based algorithms for optimizing smart bin collection in smart cities is provided in [6]. It critically examines various methodologies, including Reinforcement Learning (RL), time-series forecasting, Genetic Algorithms (GA), and Graph Neural Networks (GNNs), for their efficiency in collection processes. Huh et al. [9] introduce an IoT-based Smart Trash Bin designed to enhance recycling and waste management efficiency with reduced costs. The work highlights three innovative designs utilizing sensor, image processing, and spectroscopy technologies within an IoT framework, aiming to cut operational expenses, including labor costs. Benarbia et al. [4] present an innovative model for smart waste collection systems (SWCs) using stochastic Petri nets (PN) with inhibitor arcs and discrete event simulation. This approach addresses the challenges in waste collection by introducing a real-time inventory control system that optimizes the routing and scheduling of collection vehicles based on the fill levels of waste containers. In [3], the authors discuss a new modeling language tailored for Big Data systems to model the MapReduce paradigm. The language is designed within the SIMTHESys framework ([1,10]), enabling efficient modeling of data distribution and processing across multiple computing nodes. The deployed methodology aims to minimize the complexity of performance modeling by abstracting the underlying systems, allowing domain experts to focus on the system architecture rather than low-level details. Likotiko et al. [11] discuss an approach to optimizing waste collection through IoT technologies. The authors developed a multi-agent-based IoT architecture for monitoring and optimizing solid waste collection. Utilizing the NetLogo multi-agent platform, their system simulates real-time scenarios of waste bin fill levels and truck collection processes, enabling dynamic and smart decision-making. A novel approach named QueSAIR (Queuing System Assessment and Impact Reduction), is discussed in [13], integrating queuing theory into the reverse logistics network for effective inert construction waste management. This approach assesses queuing systems and their impacts through simulation and quantitative analysis. The case study focuses on reducing negative impacts like cost, emissions, noise pollution, extra fuel consumption, loss in productivity, and energy losses. Markov et al. [12] discuss optimizing waste collection routes, focusing on recyclable waste. The authors introduce a mixed binary linear programming model that accounts for various real-world complexities, such as heterogeneous vehicle fleets, multiple depots, and site-specific constraints. A significant contribution is the development of a local search heuristic capable of solving large instances with an optimality gap of less than 2%. With regard to the deployment of Markovian Agent Models (MAMs, see for example [5]), Gribaudo et al. [8] propose an Internet of Things (IoT)-based approach for monitoring cultural heritage sites, with a focus on the UNESCO-protected center of Matera, Italy. The system utilizes a mix of heterogeneous sensors and MAMs to monitor crowd behavior and anticipate threats to the site, aiming to prevent damage to cultural heritage. This approach stands out for its ability to adapt to real-time data, predict future scenarios, and optimize incident management strategies, marking a significant contribution to cultural heritage preservation and smart city applications. The modeling of sensor nodes using MAMs is discussed in [7]. This work compares four on-off strategies to manage power consumption, where sensors alternate between active, asleep, or failed states. A MAM assesses network performance and dependability, considering message routing and transmission between nodes based on geographic position. Finally, in [2], the authors present a modeling approach for evaluating the impact of storage allocation policies in geographically distributed, large-scale cloud architectures, based on MAMs. The related works differ in their approach and focus on waste collection optimization. For instance, Huh et al. focus on enhancing waste separation through the integration of multiple sensor types, while Benarbia et al. emphasize the use of stochastic Petri nets for optimizing the routing and scheduling of collection vehicles. Likotiko et al. take a multi-agent-based approach, which allows for real-time dynamic decision-making and citizen engagement, differing from the purely IoT-based systems. Gatti et al. distinguish their work by critically reviewing various algorithmic methodologies and their application to waste collection, offering insights into both classic and ML-based techniques. Our approach focuses on utilizing MAMs to simulate the behavior of garbage collectors and smart bins in a dynamic environment. Unlike the works mentioned above, which primarily rely on either machine learning or deterministic optimization approaches, we adopt a stochastic model to evaluate different garbage collection strategies. The key difference between our model and the multi-agent system proposed by Likotiko et al. is that we incorporate Markov-modulated processes to represent the fill levels of bins dynamically, allowing for a more precise analysis of system behavior under various collection policies.

3 Case Study and Model

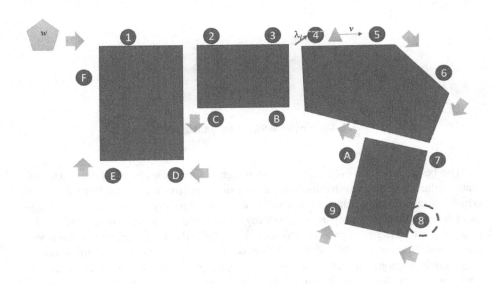

Fig. 1. The smart Waste Collection System

Figure 1 outlines a smart waste collection system composed of a base station w, a garbage collector GC, and multiple blocks. The GC is tasked with traversing clockwise a predefined circuit to collect waste from smart bins located at each block. The system's features are as follows: i) Base Station (w): The starting point for the garbage collector's route; ii) Blocks: Represented by polygons, these are the locations where smart bins are positioned; iii) Garbage Collector (GC): A mobile unit that travels at a constant speed v (meters/minute) following a clockwise direction around the circuit; iv) Smart Bins: Equipped with sensors, these bins are depicted as small circles, distributed along the blocks; v) Emptying Time (t): The time taken by the GC to empty a smart bin, measured in minutes; vi) Filling Rate: The rate at which garbage accumulates in bin i, expressed as a percentage per minute. As outlined later, we assume the filling rate follows a two-state Markov-modulated process; vii) Alarm Threshold (α): A predefined fill level at which a smart bin sends an alert to the GC, indicating an urgent need for emptying, and finally, viii) Alarm Indication: A smart bin that has surpassed the alarm threshold is marked with a dashed circle.

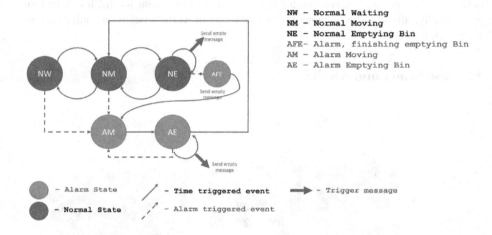

Fig. 2. Markovian Model of the garbage collector

The behavior of the GC illustrates an integrated and responsive system that optimally balances routine collection duties with the capacity to respond immediately to critical events, thereby enhancing the efficiency and reliability of urban waste management services. Incorporating a dual-modality operation underscores the system's agility in adapting to dynamic urban environments. Specifically, Fig. 2 delineates the operational modalities of the GC within a smart waste management system. Under normal operating conditions, the GC commences its cycle in a waiting state, denoted as NW, where it remains for a predetermined duration at the base w before initiating its reconnaissance. This transition is indicated by a shift from NW to the Normal Movement (NM) state, wherein the GC traverses the predefined route to service the smart bins. Subsequently, the process culminates in the Normal Emptying (NE) state, where the GC performs the waste evacuation from the smart bins as per its usual itinerary. Conversely, the system is designed to adapt to exigent circumstances necessitated by the surpassing of a predetermined fill-level threshold (α). When a smart bin reports that a threshold α has been exceeded, the GC promptly aborts its current task, irrespective of its state, and enters an Alarm Movement (AM) mode, sending alert messages. This state is characterized by the GC swiftly relocating to the smart bin that has issued the alert. Upon arrival, the GC transitions to the Alarm Emptying (AE) state to address the overfilled condition. After successfully emptying the contents of the alerted smart bin, GC can resume normal operations or react to another alarm.

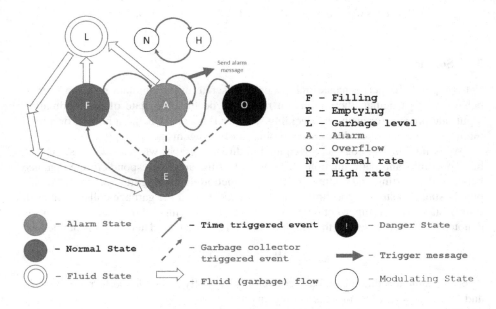

Fig. 3. Markov Model of the smart bin

Figure 3 illustrates a Markovian model of a smart bin (SB) system, where the smart bin can exist in various states. Initially, it operates in a normal state, transitioning between two primary modes: actively being filled with waste (state F) or undergoing emptying (E state) by the garbage collector. Upon surpassing a predetermined waste threshold, the smart bin's agent initiates alarm messages to the garbage collector GC agent, signaling the need for immediate emptying. However, if GC fails to promptly respond, the SB agent enters an O state, signifying the potential spillage of waste onto the street. The level of waste in the bin is represented as a continuous fluid state L, with transitions occurring towards an empty state or incoming flows either from F or A states. The flow of trash is modulated by a 2 state Continuous Time Markov Chain, that alternates between a state N with normal rate λ_N, and a state H with a high rate λ_H. In particular, it jumps from N to H at a Poisson rate γ_N, and from H to N at rate γ_H. Table 1 recaps the components of the case study, the units of measure, and the corresponding values.

Table 1. Parameters of the Smart Waste Collection System

Parameter	Unit of Measure	Value
Sweeper speed (v)	[m/min]	60
Emptying time (t)	[min]	3
Alarm threshold percentage (α)	[%]	60% ... 90%
Wait at base (w)	[min]	30
Modulation rates (γ_N, γ_H)	[hours^{-1}]	$\gamma_N = 1/3, \gamma_H = 12$
Filling rates (λ_N, λ_H)	[%$^{-1}$]	$\lambda_N = 1/5, \lambda_H = 5/2$

3.1 Solution

We compute[1] the performance indices using discrete event simulation of the agents' behavior. In particular, the simulation randomly decides the state of the modulating input rate process for each bin, deciding whether it is experiencing a normal or a high input flow rate, and then decides the evolution of the agents accordingly.

We define the topology as a weighted undirected graph, where nodes describe the bin's positions and path junctions. The weight of the arcs corresponds to the distance between the starting and ending nodes. The function describing the cycle $f_c(t)$ (computed using Dijkstra's algorithm) indicates the location of the garbage collector at each time instant, specifying whether it is moving or emptying a bin. The cycle time is denoted as t_c. Let $\mathcal{N}$ denote the graph nodes and $\mathcal{E}$ the edges. Then, $f_c(t)$ is defined as:

$$f_c : [0, t_c] \to \mathcal{N} \cup \mathcal{E} \qquad (1)$$

where $\mathcal{N}$ indicates that the garbage collector is currently at a node, performing its task, and $\mathcal{E}$ indicates that it is in transit along an edge.

The function for the time to reach the alarm $f_a(a, n)$ (computed using the minimal path) determines the time required to reach a bin that has triggered an alarm from the garbage collector's current position. $f_a(a, n)$ is defined as:

$$f_a : (\mathcal{N} \cup \mathcal{E}) \times \mathcal{N} \to \mathbb{R}^+ \qquad (2)$$

where $f_a(a, n)$ specifies the time required to reach the node $n \in \mathcal{N}$, where the alarm has been triggered, from the node $a \in \mathcal{N}$ or edge $a \in \mathcal{E}$, depending on the garbage collector's current location. The simulator uses functions $f_c(t)$ and $f_a(a, n)$ to decide the movement of the GC according to the evolution of the level of the bins, and to one of three policies defined later.

4 Experiments and Discussions

Figure 4 illustrates the evolution of the filling percentage of 4 out of 15 bins, specifically the 4 bins that have the most activity concerning alarms and potential overflows.

---

[1] The code used in this work is available at https://github.com/EBarbierato/epew2024.

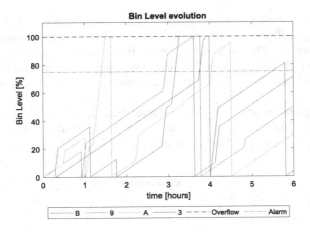

Fig. 4. The evolution of the level of the bins 3, 9, A and B.

The figure emphasises how the modulating process alters the filling rate of the bins, with two slopes: one corresponding to the slow rate, and the other to the state with a faster filling. In the same figure, it can also be observed the contribution of the garbage collector, which resets a bin to its empty state.

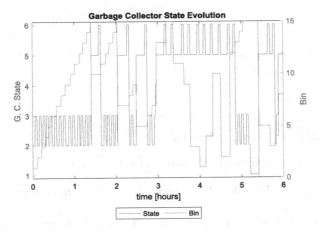

Fig. 5. The evolution of the state of the garbage collector: 1 - Waiting (NW), 2 - Moving (NM), 3 - Emptying (NE), 4 - Alarm during emptying (AFE), 5 - Alarm Moving (AM), 6 - Alarm Emptying (AE). The line on the secondary axis represents the bin currently served by the GC.

Figure 5 illustrates the evolution of the state of the GC, which alternates between movement, emptying, pause, and alarm management (blue line). State 5 is observed when the garbage collector is moving towards the bin that triggered the alarm, and state 6 is observed when it is emptying the bin that triggered the alarm. Particular attention should be given to the figure showing the agent's state at around 3 h, because at this time, the garbage collector briefly reaches state 4 due to an alarm sounding while it is emptying another bin. The same figure, on the right axis corresponding to the orange line, it shows which bin the garbage collector is going to empty. The numbers representing the bins range from 1 to 15 (with a range of $10 - 15$ corresponding to bins $A - F$).

Figure 6 depicts the state of the 4 selected bins and shows how they alternate between normal filling, alarm state, overflow state, and finally emptying. Note how the emptying states correspond to the same action performed by the CG agent.

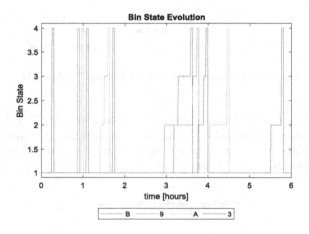

Fig. 6. The evolution of the state of the bins 3, 9, A and B: 1 - Filling (F), 2 - Alarm (A), 3 - Overflow (O), 4 - Emptying (E).

Considered Garbage Collection Policies

The following policies have been taken into account:

1. *Ignore* - The alarms from the smart bins are ignored. The route is followed as if the garbage bins were not smart.
2. *Continue* - When an alarm is triggered, the garbage collector immediately proceeds to empty the bin that raised the alarm as soon as possible. After emptying it, if some other alarms or overflows occur in the meantime, it goes to address those. Priority is given to overflows: if any bin has overflowed during the emptying process, the GC first deals with those before attending to other alarms. After emptying all problematic bins, the route continues from the point reached by the GC.
3. *Resume* - As the *Continue* policy, but in this case when the GC returns to normal operation, the inspection route resumes from where it was interrupted.

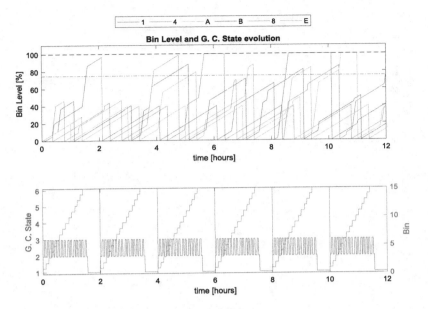

Fig. 7. The *Ignore* policy for 6 randomly selected nodes 1, 4, A, B, 8, and E. Bin occupancy status is at the top; the agent's status is at the bottom.

Figure 7 focuses on the *Ignore* policy, and it shows that the agent ignores all alarms and only alternates between the states of movement, emptying, and pause. It illustrates how this affects six bins that experience more overflows and alarms, suggesting that these bins operate autonomously without any interaction with the agent.

The *Continue* policy is depicted in Fig. 8. The GC, after responding to an alarm, continues its route from the point where it had arrived due to the alarm. Essentially, during its route, if it moves towards a bin and skips others, it then ignores the skipped bins and continues from where it went due to the alarm. Alternatively, if the alarm causes the agent to backtrack, it resumes and visits the same bins multiple times. In this scenario, it can be observed that when a bin triggers an alarm, the agent moves to and empties it.

The *Resume* policy is shown in Fig. 9. It can be observed, when looking at the orange line representing the bin the agent is emptying, that after jumping up and down to manage exceptions, it resumes from where it had last stopped.

Evaluation

Initially, we consider an alarm threshold set at 75%. Later we will study the impact of the threshold on the performance of the system. Figure 10 shows the average number of times one bin is cleaned according to each policy, and emphasises how the bins are handled differently. Bins located closer to the start of the first route are emptied at a time when they cannot yet be filled because they are emptied almost immediately. One day is insufficient for the system to reach a steady state, and the transient caused by the

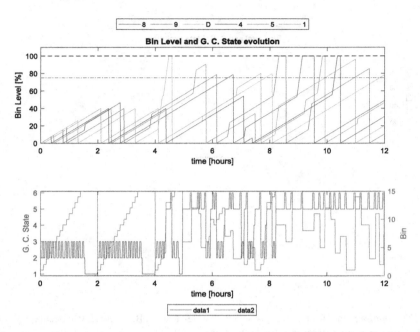

Fig. 8. The *Continue* policy for 6 randomly selected nodes 8, 9, D, 4, 5, and 1. Bin occupancy status is at the top; the agent's status is at the bottom.

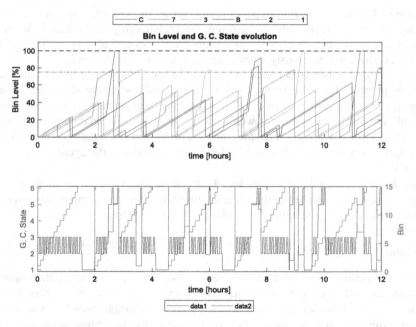

Fig. 9. The *Resume* policy for 6 randomly selected nodes C, 7, 3, B, 2, and 1. Bin occupancy status is at the top; the agent's status is at the bottom.

bin from where the route begins has a very different impact, which also depends on the three policies.

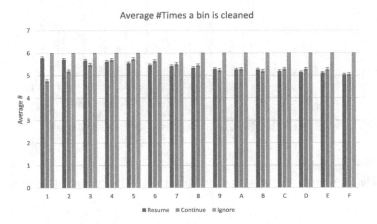

Fig. 10. Average number of cleaning per bin (including the confidence intervals), for the considered policies.

Figure 11 shows the time that elapses between the emptying of one bin and another. In this case, even if the alarm is ignored, the average time between two emptying is not constant, even for the policy that ignores alarms, because some bins are initially emptied even though there has not been enough time for them to fill up. Although the considered cleaning cycle time is two hours, it is evident how the policies lead to varying periods of time, either longer or shorter, depending on the position of the bins.

Figures 12, 13 and 14 focus on examining the importance of the threshold. It considers the three policies by setting three alarm levels (60%, 75%, and 90%). Figure 12 shows that the policy that ignores the threshold results in the fewest alarms, while the other two have almost identical results. Even if the *Ignore* policy does not consider the level of bins, a smaller threshold causes a larger number of alarms. Figure 13 shows that the *Ignore* policy has also the smallest number of overflows per day, unless the threshold is very high, in the case in which the *Resume* policy performs better. However, as shown in Fig. 14, the *Ignore* policy is also the one that has the longest time during which a bin remains in overflow, which represents a critical moment for the system.

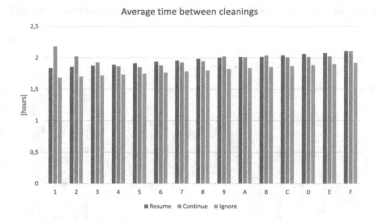

Fig. 11. Average time between cleanings per bin, for the considered policies.

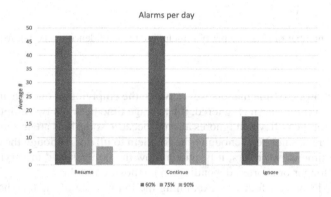

Fig. 12. Average number of alarms per day, for the considered policies and different thresholds.

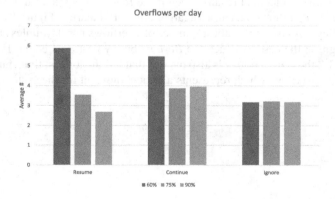

Fig. 13. Average number of overflows per day, for the considered policies and different thresholds.

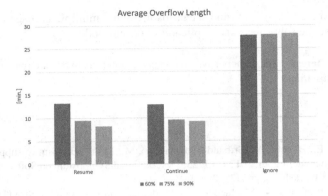

Fig. 14. Average duration of an overflow, for the considered policies and different thresholds.

Results

This analysis concludes that the optimal approach is to handle the alarm, but then resume the route to avoid the risk that, over time, the last bins are always and only managed due to alarms. This situation could overwhelm the system. Resuming the route ensures that when the alarms are limited, they do not disrupt the system or cause the garbage collector to jump from one alarm to another, but instead allow for preventive emptying.

The system performs better with a higher threshold (90%). The threshold should be set as high as possible so that the agent can reach the bin within the short time between when the alarm is triggered and when it reaches the bin that sent the alarm. If the alarm is triggered too early, the system becomes unstable. However, if the garbage collector can reach the bin that raised the alarm just before it overflows, it can effectively solve the situation, without affecting too much the regular operation cycle. This approach stabilizes the system as if there were no alarms, thanks to the fact that the garbage collector resumes its route from where it was interrupted and remains responsive to incoming alarms.

5 Conclusions

In this study, we addressed the challenge of optimizing garbage collection in environments equipped with smart bins. By developing a Markovian Agent Model, we analyzed various collection strategies and their thresholds, particularly focusing on the behaviors and interactions between the garbage collector and the smart bins. Our findings indicate that actively responding to alarms and resuming the predetermined route post-response (Resume policy) offers an optimal balance, ensuring bins are serviced efficiently without overwhelming the system. Moreover, setting higher alarm thresholds, such as 90%, proved beneficial, allowing the garbage collector sufficient time to address alerts without frequent disruptions. This approach not only stabilizes the system but also minimizes instances of overflow. Future work will delve into more complex topologies, diverse filling rates, and scenarios involving multiple garbage collectors to further enhance the robustness and applicability of our model.

Disclosure of Interests. The authors have no competing interests to declare that are relevant to the content of this article.

References

1. Barbierato, E., Bobbio, A., Gribaudo, M., Iacono, M.: Multiformalism to support software rejuvenation modeling. In: 2012 IEEE 23rd International Symposium on Software Reliability Engineering Workshops, pp. 271–276. IEEE (2012)
2. Barbierato, E., Gribaudo, M., Iacono, M.: Modeling hybrid systems in SIMTHESys. Electron. Theor. Comput. Sci. **327**, 5–25 (2016)
3. Barbierato, E., Gribaudo, M., Iacono, M., et al.: A performance modeling language for big data architectures. In: ECMS, pp. 511–517 (2013)
4. Benarbia, T., Darcherif, A.M., Sun, D.J.: Modelling and performance analysis of smart waste collection system: a petri nets and discrete event simulation approach. Int. J. Decision Supp. Syst. **4**(1), 18–40 (2019)
5. Bobbio, A., Cerotti, D., Gribaudo, M., Iacono, M., Manini, D.: Markovian agent models: a dynamic population of interdependent markovian agents. In: Al-Begain, K., Bargiela, A. (eds.) Seminal Contributions to Modelling and Simulation, pp. 185–203. Springer International Publishing, Cham (2016). https://doi.org/10.1007/978-3-319-33786-9_13
6. Gatti, A., Barbierato, E., Pozzi, A.: Toward greener smart cities: A critical review of classic and machine-learning-based algorithms for smart bin collection. Electronics **13**(5) (2024). https://doi.org/10.3390/electronics13050836, https://www.mdpi.com/2079-9292/13/5/836
7. Gribaudo, M., Cerotti, D., Bobbio, A.: Analysis of on-off policies in sensor networks using interacting Markovian agents. In: 2008 Sixth Annual IEEE International Conference on Pervasive Computing and Communications (PerCom), pp. 300–305. IEEE (2008)
8. Gribaudo, M., Iacono, M., Levis, A.H.: An iot-based monitoring approach for cultural heritage sites: The matera case. Concurr. Comput.: Pract. Exp. **29**(11), e4153 (2017)
9. Huh, J.H., Choi, J.H., Seo, K.: Smart trash bin model design and future for smart city. Appl. Sci. **11**(11) (2021). https://doi.org/10.3390/app11114810, https://www.mdpi.com/2076-3417/11/11/4810
10. Iacono, M., Barbierato, E., Gribaudo, M.: The simthesys multiformalism modeling framework. Comput. Math. Appl. **64**(12), 3828–3839 (2012). https://doi.org/10.1016/J.CAMWA.2012.03.009, https://doi.org/10.1016/j.camwa.2012.03.009
11. Likotiko, E.D., Nyambo, D., Mwangoka, J.: Multi-agent based iot smart waste monitoring and collection architecture. arXiv preprint arXiv:1711.03966 (2017)
12. Markov, I., Varone, S., Bierlaire, M.: Vehicle routing for a complex waste collection problem. In: 14th Swiss Transport Research Conference (2014)
13. Zhang, X., Ahmed, R.R.: A queuing system for inert construction waste management on a reverse logistics network. Autom. Constr. **137**, 104221 (2022)

Approximation of First Passage Time Distributions of Compositions of Independent Markov Chains

András Horváth[1]($\boxtimes$)[iD], Marco Paolieri[2][iD], and Enrico Vicario[3][iD]

[1] Department of Computer Science, University of Turin, Turin, Italy
horvath@di.unito.it
[2] Department of Computer Science, University of Southern California, Los Angeles, USA
paolieri@usc.edu
[3] Department of Information Engineering, University of Florence, Florence, Italy
enrico.vicario@unifi.it

Abstract. To improve performance or reliability, systems frequently include multiple components that operate in parallel or with limited interaction, e.g., replicated components for triple modular redundancy. We consider components modeled by independent and possibly different continuous-time Markov chains and propose an approach to estimate the distribution of first passage times for a combination of component states (e.g., a system state where all components have failed) without generating the joint state space of the underlying Markov chain nor evaluating probabilities for each of its states. Our results highlight that the approach leads to accurate approximations with significant reductions of computational complexity.

Keywords: First Passage · Bounded Reachability · CTMC · Markov Chain · Replicated · Modular Redundancy

1 Introduction

Continuous-time Markov chains (CTMCs) are a class of stochastic processes that has found broad application in models of system performance and reliability. Many high-level modeling formalisms such as stochastic Petri nets [17], stochastic process algebras [9], and queuing networks [15] define CTMC processes that can then be analyzed using dedicated tools [1,12,16] to evaluate transient or steady-state metrics.

To improve performance or reliability, systems frequently include replicated components operating in parallel or with limited interaction, e.g., components replicated for triple modular redundancy (TMR). When the system includes many replicated components, each with a state evolving over time, the large number of states of the resulting CTMC process presents major challenges due to memory and computation requirements. These issues are exacerbated by the necessity of using phase-type (PH) distributions [18] to model activities with non-exponential durations. PH distributions can be introduced in the model as a sequence of intermediate states (phases) for a non-exponential activity, where rates between states are selected to fit the original distribution (by matching moments [4], tail behavior [11], or both [13], or by maximizing likelihood [3]). While the use of additional intermediate states allows more accurate approximations, it also increases the number of system states, especially when many such activities can execute in parallel.

Several approaches were proposed to analyze large CTMCs with multiple components. These include *binary decision diagrams* (BDDs) [16], which represent the CTMC transition matrix as a directed graph where paths select rate values, and *structured analysis* [6], which represents the CTMC transition matrix as a sum of Kronecker products of the small transition matrices of individual components. Either approach can be used to evaluate transient or steady-state probabilities of each system state through iterative methods.

In this work, we propose an *approximate solution method* to evaluate the cumulative distribution function (cdf) of the first passage time of a combination of component states (e.g., a system state where all components have failed) *without enumerating system states* (in contrast with alternative approaches computing probabilities for each system state). The approximation will be developed by time discretization assuming that the system can be in the concerned combination of component states at the end of a small time interval only if all but one of the components are already there at the beginning of the time interval. The state distribution of the components at the beginning of each interval is assumed to be independent, irrespective of first passage events in previous time intervals. Time discretization will be removed then leading to a differential equation whose numerical integration provides the approximation.

In contrast with classical work on replicated components [2,5,19], we consider possibly different CTMC components with an arbitrary number of states (instead of one up and one down state) and evaluate the cdf (instead of the moments) of the first passage time.

The paper is organized as follows. We recall background and define our problem in Sect. 2, and we present our approximation method in Sect. 3. In Sect. 4, we provide numerical results and discuss some implementation issues, drawing our conclusions in Sect. 5.

2 Background and Problem Definition

We consider a system composed of n independent components modeled as CTMCs $\{Y_k(t)\}_{1 \leq k \leq n}$ with finite state spaces S_k and infinitesimal generators $Q_k = (q_{kij})_{i,j \in S_k}$ with $k = 1, ..., n$. We denote by $F_k \subseteq S_k$ the set of failure states of component k. In general, the components can be repaired (i.e., the states in F_k are not absorbing). The transient probabilities of component k are denoted by

$$p_{kij}(t) = P(Y_k(t) = j \mid Y_k(0) = i) \text{ for } i, j \in S_k, k = 1, ..., n,$$

which can be calculated, for example, by uniformization.

The system can be modeled as a CTMC $\{X(t)\}$ with state space denoted by $S = S_1 \times S_2 \times ... \times S_n$, where $X(t) = (Y_1(t), \ldots, Y_n(t))$. The set of states in which all components are in a failure state is denoted by $F = F_1 \times F_2 \times ... \times F_n \subseteq S$.

Let T be the first time when all the components are in a failure state, that is,
$$T = \min\{t \geq 0 \mid X(t) \in F\},$$
and let $F_T(t)$ indicate its cdf, that is, $F_T(t) = P(T \leq t)$. The cdf $F_T(t)$ is known as *first passage time distribution* to reach a state in F. It is also referred to as *time-bounded reachability* in probabilistic model checking.

Since the components are independent, the transient probabilities of the composed system can easily be computed based on the transient probabilities of the components in product form as

$$P(X(t) = (y_1, ..., y_n) \mid X(0) = (s_1, ..., s_n)) = \prod_{k=1}^{n} p_{k s_k y_k}(t). \qquad (1)$$

where $(s_1, ..., s_n)$ is the initial state. Accordingly, also $P(X(t) \in F)$, that is, the probability that all components are in a failure state at a given time t (which is not equal to $F_T(t)$), can be computed in product form considering the components in isolation.

On the contrary, $F_T(t)$, i.e., the first passage time cdf to reach a state in F, cannot be obtained in product form multiplying first passage time cdfs of the components. Indeed, the product of first passage time cdfs of the components yields the probability that all components have been in a failure state before time t at least once (and not the probability that all components have been in a failure state at the same time).

In order to determine $F_T(t)$ exactly, one has to consider a variant of $\{X(t)\}$, that we denote by $\{X'(t)\}$, in which states in F are made absorbing. In this modified CTMC we have

$$F_T(t) = P(X'(t) \in F \mid X'(0) = (s_1, ..., s_n)).$$

Making states in F absorbing couples the behavior of the components and they are not independent anymore in a probabilistic sense. Consequently, the transient probabilities of $\{X'(t)\}$ are not in product form and we need to consider a CTMC with $|S| - |F|$ states (it is not necessary to represent states in F during the calculations), with exponential growth of the state space with respect to the number of components, quickly making the analysis unfeasible.

The aim of this paper is to propose an approximation of $F_T(t)$, denoted by $\hat{F}_T(t)$, that is based on the individual behavior of the components and hence does not require analyzing $\{X'(t)\}$. By doing so, the computational complexity is kept linear with respect to the number of components.

3 Approximation Method

In order to provide the stochastic interpretation of the proposed approximation method, we give first a description in which time is discretized. This discretized version proceeds in time by taking steps of length δ and calculating $\hat{F}_T(i\delta), i = 0, 1, 2, \ldots$, the approximation of $F_T(i\delta)$ for $i = 0, 1, 2, \ldots$. We assume that δ is such that there is negligible probability that more than one component makes a transition in an interval of length δ.

The assumption on δ implies that the system enters a state in F in $(t, t+\delta]$ only if at time t the number of failed components is $n - 1$. For this reason, at each step we calculate the probability that in $\{X(t)\}$ at time t all components other than component k are failed by

$$U_k(t) = \prod_{1 \leq i \leq n, i \neq k} P(Y_i(t) \in F_i). \qquad (2)$$

The intensity with which component k moves from up states to down states at time t can be calculated as

$$D_k(t) = \sum_{i \notin F_k} \left(P(Y_k(t) = i) \sum_{j \in F_k} q_{kij} \right). \qquad (3)$$

The probability itself that component k is up at time t and makes a transition from an up state to a down state in $(t, t+\delta]$ can be approximated by $D_k(t)\delta$.

In order to easily consider all components together, we introduce also

$$G(t) = \sum_{i=1}^{n} U_i(t) D_i(t). \qquad (4)$$

The approximation starts with $\hat{F}_T(0) = F_T(0) = P(X(0) \in F)$, which can be easily calculated given the initial distribution of the components, and proceeds according to

$$\hat{F}_T((i+1)\delta) = \hat{F}_T(i\delta) + (1 - \hat{F}_T(i\delta)) G(i\delta)\delta, \qquad (5)$$

where we multiply by $(1 - \hat{F}_T(i\delta))$ to consider the probability that a state in F was already reached.

The approximation in Eq. (5) relies on the fact that in $\{X'(t)\}$ the components evolve independently up to the moment in which the system reaches a state in F. This suggests that the product-form probabilities in Eq. (1) multiplied by the probability that a state in F has not been reached give a good approximation of the probabilities of the states in $S \setminus F$, that is,

$$P(X'(t) = (y_1, \ldots, y_n)) \approx (1 - F_T(t)) \prod_{i=1}^{n} P(Y_i(t) = y_i) \text{ for } (y_1, \ldots, y_n) \notin F,$$

where in our numerical scheme $F_T(t)$ is substituted by $\hat{F}_T(t)$. For the states in F we have $P(X'(t) \in F) = F_T(t) \approx \hat{F}_T(t)$.

To remove time discretization, note that Eq. (5) can be reorganized as

$$\frac{\hat{F}_T((i+1)\delta) - \hat{F}_T(i\delta)}{\delta} = (1 - \hat{F}_T(i\delta)) G(i\delta),$$

and by taking the limit $\delta \to 0$, we obtain the differential equation

$$\hat{F}_T'(t) = (1 - \hat{F}_T(t))G(t) \qquad (6)$$

with initial condition $\hat{F}_T(0) = F_T(0)$. The approximation $\hat{F}(t)$ can then be calculated by numerical integration of Eq. (6).

Note that if the components are identical and share also the initial distribution, then

$$U(t) := U_1(t) = ... = U_n(t) = P(Y_1(t) \in F_1)^{n-1},$$

$$D(t) := D_1(t) = ... = D_n(t) = \sum_{i \notin F_1} \left(P(Y_1(t) = i) \sum_{j \in F_1} q_{1ij} \right),$$

and Eq. (4) simplifies to $G(t) = nU(t)D(t)$.

4 Numerical Experiments and Implementation Issues

We present two sets of experiments. First, we use a model composed of identical components described by a CTMC with a small state space. This case allows us to compare the results obtained by the proposed approximation method against exact results. Next, a model composed of different CTMCs with large state spaces is analyzed. In this case we compare the approximation against simulation.

After the experiments, we briefly discuss implementation issues and provide execution times.

4.1 Identical Components with Small State Space

The infinitesimal generator of the components is

$$Q = \begin{pmatrix} -\alpha & \alpha & 0 & 0 \\ \alpha & -2\alpha - \beta & \alpha & \beta \\ 0 & \alpha & -\alpha - \beta & \beta \\ \gamma & 0 & 0 & -\gamma \end{pmatrix},$$

which describes a four state system in which (i) the first three states correspond to normal operation states among which there are transitions with intensity α, (ii) from the second and third up states, the down state can be reached by a transition with intensity β, and (iii) repair takes to the first state with intensity γ.

The parameters (α, β, γ) allow us to calibrate the steady-state probability of being in the failure state in a component considered in isolation which, as we will see, has an impact on the accuracy of the approximations. The following sets of parameters (α, β, γ) will be used: $(1, 1, 12)$ (Case 1), $(2, 2, 3)$ (Case 2), $(4, 4, 0.75)$ (Case 3). Steady-state probabilities of the failure state are $\frac{1}{33}, \frac{1}{5}, \frac{2}{3}$, respectively.

We calculated exact and approximate first passage distributions, i.e., $F_T(t)$ and $\hat{F}_T(t)$, for several values of n in the above cases. The initial state s is the state in which all components are in the first state.

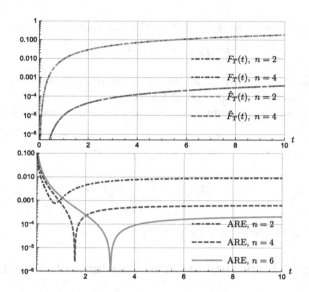

Fig. 1. Case 1: Exact $F_T(t)$ and approximate $\hat{F}_T(t)$ (top); approximation ARE (bottom).

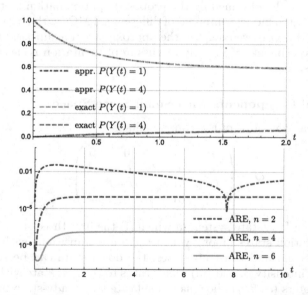

Fig. 2. Case 1: Approximate and exact state probabilities with $n = 2$ (top); ARE of the probability of the fourth state (i.e., the failure state) for different values of n (bottom).

Figure 1 shows $F_T(t)$ and $\hat{F}_T(t)$ and the absolute relative error (ARE) of $\hat{F}_T(t)$. Visually, the approximation cannot be distinguished from the exact values. The ARE shows that the approximation error is low and decreases as we increase the number of components. The downward spikes in the ARE are due to the points where exact and approximate curves cross each other leading to a point where the ARE is zero. In order to investigate the source of the approximation error of $\hat{F}_T(t)$, we calculated state probabilities of the components based on the approximation and exactly. In Fig. 2 there is no visible error in the state probabilities with $n = 2$ and the ARE of the probability of the fourth state 4 decreases as n is increased. (Note that since the initial state is the same for all components, they have the same transient probabilities; hence we use $Y(t)$ without specifying the component.)

In Fig. 3, we observe that for Case 3, where the failure state is reached with higher probability, we obtain greater absolute error and ARE of the approximation. Similarly to Fig. 1, the error decreases as we increase the number of components. In Fig. 4, we can see that (as expected) the errors in the state probabilities and ARE of the failure state are also higher.

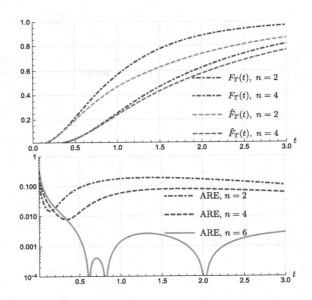

Fig. 3. Case 3: Exact $F_T(t)$ and approximate $\hat{F}_T(t)$ (top); approximation ARE (bottom).

As a comparison of the three cases, in Fig. 5, we depicted for $n = 4$ the ARE of $\hat{F}_T(t)$ and the ARE of the probability of the fourth state. In the range $t \in [0, 2]$, as expected, the higher the probability of the failure state, the worse approximation we obtain. For the third considered case, after $t = 2$ the approximation gets better. This is due to the fact that $F_T(t)$ is getting closer and closer to one making it easier to obtain good ARE values. In the other two cases, we have $F_T(10) < 0.2$ and the approximation error remains stable.

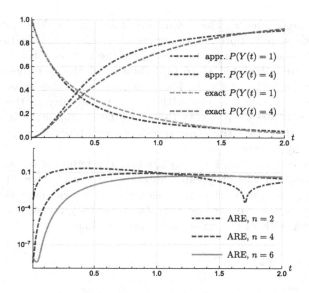

Fig. 4. Case 3: Approximate and exact state probabilities with $n = 2$ (top); ARE of the probability of the fourth state (i.e., the failure state) for different values of n (bottom).

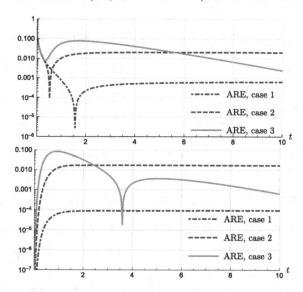

Fig. 5. ARE of $\hat{F}_T(t)$ (top) and ARE of the probability of the fourth state (bottom) with $n = 4$ for all three considered cases.

4.2 Different Components with Large State Spaces

In order to describe the components, we use the Petri net (PN) depicted in Fig. 6 which models a rejuvenation mechanism [8]. The system is in one of four states:

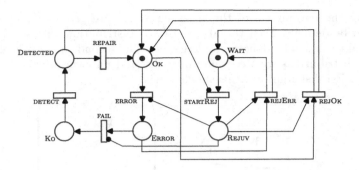

Fig. 6. PN model of a rejuvenation mechanism.

OK is a safe operational state; ERROR corresponds to an aged operational state that may lead to a failure; in state KO the system is down due to a failure; the fact that a failure has been detected is modeled by state DETECTED. Transitions among these four states are modeled by PN transitions ERROR, FAIL, DETECT, and REPAIR. The system is complemented by a rejuvenation mechanism with a timer that controls when the next rejuvenation takes place. The rejuvenation mechanism is either in state WAIT, where it waits for the timer to run out, or in state REJUV, where rejuvenation is carried out. The transition from WAIT to REJUV is called STARTREJ. Rejuvenation can be completed while the system is in state OK or ERROR and it takes the system back to its initial state through either transition REJOK or REJERR, respectively. In state KO rejuvenation is not possible but the timer is stopped only when a failure is detected. When rejuvenation is in progress, the system cannot degrade, that is, transitions ERROR and FAIL are inhibited when place REJUV has a token.

The time to fire distribution of the transitions will be defined through PH distributions [18]. An order r PH distribution is given through the time to absorption in a CTMC with r transient states, called phases, and one absorbing state. Accordingly, it is determined by the initial probability vector, denoted by a, and the infinitesimal generator of its CTMC, denoted by A. We will use two subclasses of PH distributions. The first one is the family of Erlang distributions. In terms of PH distributions, an Erlang distribution with shape parameter[1] r and mean equal to m is obtained by

$$a = (1\ 0\ \ldots\ 0),\ A = \begin{pmatrix} -\frac{r}{m} & \frac{r}{m} & & & \\ & -\frac{r}{m} & \frac{r}{m} & & \\ & & \ddots & \ddots & \\ & & & -\frac{r}{m} & \frac{r}{m} \\ & & & & -\frac{r}{m} \end{pmatrix},$$

---

[1] The shape parameter is equal to the number of phases, that is, the number of transient states.

where only the parameters of the transient states are given in a and A (the others can be deduced from these). An Erlang distribution with r phases and mean equal to m will be denoted by $\mathrm{Erl}(r,m)$. The second subclass is a mixture of Erlang distributions with common intensity and uniform mixing probability. In terms of PH distributions, this subclass has

$$a = \left(\underbrace{\frac{1}{k}\ \frac{1}{k}\ \cdots\ \frac{1}{k}}_{k}\ \underbrace{0\ \ldots\ 0}_{r-k} \right),\ A = \begin{pmatrix} -\lambda & \lambda & & & \\ & -\lambda & \lambda & & \\ & & \ddots & \ddots & \\ & & & -\lambda & \lambda \\ & & & & -\lambda \end{pmatrix},$$

which are determined by three parameters: the total number of phases, r, the number of phases with non-zero initial probability, k, and the intensities in A denoted by λ. We will refer to this family of distributions as Erlang mixture (EM); an EM distribution will be denoted by $\mathrm{EM}(r,k,\lambda)$. In Fig. 7 we show a few examples of Erlang and EM distributions through their probability density functions (pdf).

When time to fire distributions in a PN are PH distributions, the underlying stochastic process is a CTMC whose infinitesimal generator can be built by Kronecker operations (see, e.g., [7]). The CTMC is subject to the so-called state space explosion problem. This is due to the fact that, given a marking, the number of states corresponding to the marking in the underlying CTMC is equal to the product of the number of phases of the PH distributions of the enabled transitions in the marking.

As Case 1, we consider a system composed of three components. Their PH distributions together with the number of states of the resulting CTMC are reported in Table 1. Note that the components are not identical. Every state of the CTMC corresponding to a marking in which there is a token in place KO or DETECTED is considered as a failure state. These markings are (KO,WAIT), (KO,REJUV), (DETECTED,WAIT), and (DETECTED,REJUV). For example, for what concerns the second component, the above four markings correspond to 600,

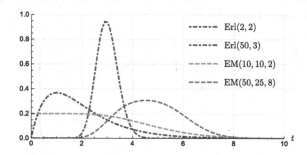

Fig. 7. Pdf of Erlang and EM distributions with various parameters.

Table 1. PH distributions of the components of Case 1, the number of states of the resulting CTMC and the number of up states in the CTMC.

	ERROR	FAIL	DETECT	REPAIR	STARTREJ	REJERR	REJOK	# st.	# up st.
comp. 1	Erl(2,40)	Erl(2,50)	EM(4,2,1)	EM(40,20,1)	Erl(100,60)	EM(8,6,1)	EM(6,3,1)	898	414
comp. 2	Erl(2,40)	Erl(2,25)	EM(4,2,1)	EM(40,20,1)	Erl(150,50)	EM(8,6,2)	EM(6,3,1)	1298	614
comp. 3	Erl(2,40)	Erl(2,50)	EM(4,2,2)	EM(40,20,2)	Erl(100,50)	EM(8,6,2)	EM(6,3,1)	898	414

4, 40 and 40 states, respectively, for a total of 684 failure states. In Table 1 we reported also the number of up states for each component. The product of these numbers provides the number of states of the CTMC that we should analyze in order to calculate the first passage time distribution exactly. For Case 1, it is 105,237,144 (i.e., approximately 10^8).

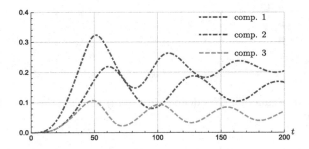

Fig. 8. Total probability of failure states as function of time for the three components of Case 1 in isolation.

To give an idea of the behavior of the components in isolation, in Fig. 8 we show the probability of being in a failure state as function of time for the three components.

The approximate first passage time distribution calculated by the proposed method is depicted in Fig. 9 together with an empirical cdf obtained by simulation. The number of simulation runs was set to $5 \cdot 10^5$. The figure also shows the 95% confidence band constructed around the empirical, simulation-based cdf using the Dvoretzky-Kiefer-Wolfowitz inequality [10,21]. The approximate first passage time distribution cannot be distinguished from the one obtained by simulation. The width of the confidence band is 0.00384. Figure 10 shows the function $G(t)$ defined in Eq. (4) which is used to obtain the approximation $\hat{F}_T(t)$ through numerical integration of Eq. (6).

As Case 2, we consider a system obtained by small modifications of the system of Case 1 in such a way that failure probabilities become smaller. Specifically, the time to fire distribution of transition REPAIR of components 1 and 2 of Case 2 is EM(40,20,5) (instead of EM(40,20,1)) which means that these two components get back to state OK from state DETECTED five times faster. This change does

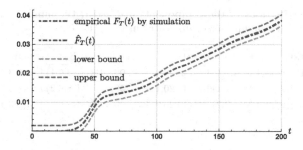

Fig. 9. Case 1: First passage time distribution by approximation and by simulation with 95% confidence bounds based on simulation.

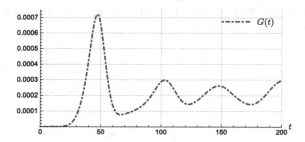

Fig. 10. Case 1: The function $G(t)$ defined in Eq. (4) which is the base of the approximation procedure.

not modify the state space, that is, the number of states and the number of up states is the same as in Case 1, reported in Table 1. The approximate first passage time distribution together with results obtained by 10^6 simulation runs are provided in Fig. 11. The approximate cdf deviates very slightly from the simulation-based cdf from about $t = 150$. Note however that, as indicated by the confidence band whose width is 0.00272, a much larger number of simulation runs would be necessary to estimate such small probabilities with high confidence. That is, the visible but very small difference can be due to chance.

Case 3 is obtained from Case 2 by increasing the number of phases of the applied Erlang distributions. Specifically, we double the number of phases of transition WAIT and change the number of phases of transitions ERROR and FAIL to 5 (in the two cases before it was 2). These modifications have a twofold impact. First, failure probabilities become even smaller because the Erlang distribution with shape parameter equal to 5 has a smaller variability and it becomes less likely that transitions ERROR and FAIL fire before a rejuvenation. Second, the state space becomes larger. The number of states of the three components is 2,898, 4,298 and 2,898, respectively. The number of up states in the components is 2,014, 3,014 and 2,014, respectively, meaning that exact analysis would require dealing with a CTMC with $12, 225, 374, 744 \approx 1.2 \cdot 10^{10}$ states. Results are shown in Fig. 12 with $2 \cdot 10^6$ simulations runs. Also in this case there is a small but

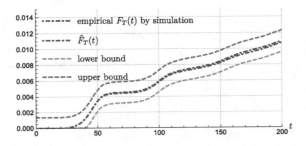

Fig. 11. Case 2: First passage time distribution by approximation and by simulation with 95% confidence bounds based on simulation.

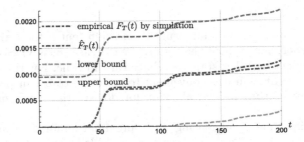

Fig. 12. Case 3: First passage time distribution by approximation and by simulation with 95% confidence bounds based on simulation.

visible difference between the approximation and the simulation-based empirical cdf but, as indicated by the confidence band with width equal to 0.00192, it can very well be due to chance and only an extremely large number of simulation traces could verify the precision of the approximation.

The intended use of the proposed approximation is the analysis of reliability of systems composed of independent components. Consequently, we are interested in approximating relatively small probabilities. In Fig. 13 we show that the approximation can result in good precision even in case of much larger probabilities by evaluating Case 1 up to $t = 10000$.

4.3 Implementation Issues

The presented numerical results were calculated by a prototype implementation of the method using Wolfram Mathematica [14]. The transient probabilities of the components, which are necessary to compute $G(t)$ defined in Eq. (4), were determined by uniformization (see, e.g., [20]) with precision 10^{-8}, representing the infinitesimal generators of the components by sparse matrices. The approximate first passage time distribution was calculated by numerical integration of the differential equation Eq. (6) applying the NDSOLVE function of Mathematica. The sought relative precision was set to 10^{-8}. NDSOLVE evaluates $G(t)$ at several time points in order to compute $\hat{F}_T(t)$. Calculation of the transient prob-

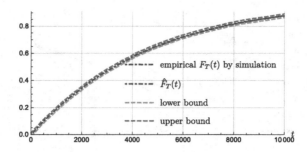

Fig. 13. Case 1: First passage time distribution by approximation and by simulation with 95% confidence bounds based on simulation up to $t = 10000$.

abilities of the components by uniformization is efficient if these time point are in increasing order. This is however not always the case since NDSOLVE, in order to guarantee precision, takes steps also backward in time. For this reason, for a time-efficient computation it is convenient to store the transient probabilities at a few recently used time points, allowing for not starting uniformization from $t = 0$.

The execution time of calculating the approximation for the complex system referred to as Case 1 in Sect. 4.2 (described in Table 1) up to $t = 200$ took 0.7 s. Case 2 required the same amount of time. Case 3 required instead about 4 s. This is because in this system both the state spaces and the intensities in the infinitesimal generators are larger (meaning that more steps are required for the uniformization to guarantee the same precision).

Simulation was also carried out in Mathematica based on the Petri net (that is, not the underlying CTMC) generating firing times according to the PH distributions of the transitions. A standard laptop was used parallelizing the generation of the simulation traces among 14 processor cores. Generating the $5 \cdot 10^5$ traces for Case 1 required around 30 min of computation. Case 2 required about an hour because generating a trace requires roughly the same amount of time but we generated twice as many traces. Simulation of Case 3 took about 5 h because a single trace requires more time and we also generated more traces.

5 Conclusions

We presented an approximate solution to compute the cdf of the first passage time of a combination of component states without enumerating system states. The approximation achieves high accuracy when failures are rare or when the system includes many components. In future work, we plan to extend the approach to analyze systems with m-out-of-n failure conditions. We also plan to investigate how far $\{X'(t)\}$ is from a product form.

References

1. Amparore, E.G.: Stochastic modelling and evaluation using GreatSPN. SIGMETRICS Perform. Evaluation Rev. **49**(4), 87–91 (2022)
2. Barlow, R.E., Proschan, F.: Theory of maintained systems: distribution of time to first system failure. Math. Oper. Res. **1**(1), 32–42 (1976)
3. Bobbio, A., Cumani, A.: ML estimation of the parameters of a PH distribution in triangular canonical form. Comp. Perf. Eval. **22**, 33–46 (1992)
4. Bobbio, A., Horváth, A., Telek, M.: Matching three moments with minimal acyclic phase type distributions. Stoch. Models **21**, 303–326 (2005)
5. Brown, M.: The first passage time distribution for a parallel exponential system with repair. In: Proceedings of the Conference on Reliability and Fault Tree Analysis, pp. 365–396. Society for Industrial and Applied Mathematics (1975)
6. Buchholz, P., Katoen, J., Kemper, P., Tepper, C.: Model-checking large structured Markov chains. J. Log. Algebraic Methods Program. **56**(1–2), 69–97 (2003)
7. Buchholz, P., Kemper, P.: Kronecker based matrix representations for large Markov models. In: Baier, C., Haverkort, B.R., Hermanns, H., Katoen, J.-P., Siegle, M. (eds.) Validation of Stochastic Systems. LNCS, vol. 2925, pp. 256–295. Springer, Heidelberg (2004). https://doi.org/10.1007/978-3-540-24611-4_8
8. Carnevali, L., Paolieri, M., Reali, R., Scommegna, L., Vicario, E.: A Markov regenerative model of software rejuvenation beyond the enabling restriction. In: Proceedings of IEEE ISSRE Workshops (WOSAR), pp. 138–145. IEEE (2022)
9. Clark, A., Gilmore, S., Hillston, J., Tribastone, M.: Stochastic process algebras. In: Bernardo, M., Hillston, J. (eds.) SFM 2007. LNCS, vol. 4486, pp. 132–179. Springer, Heidelberg (2007). https://doi.org/10.1007/978-3-540-72522-0_4
10. Dvoretzky, A., Kiefer, J., Wolfowitz, J.: Asymptotic minimax character of the sample distribution function and of the classical multinomial estimator. Ann. Math. Stat. **27**(3), 642–669 (1956)
11. Feldman, A., Whitt, W.: Fitting mixtures of exponentials to long-tail distributions to analyze network performance models. Perf. Eval. **31**, 245–279 (1998)
12. Hensel, C., Junges, S., Katoen, J., Quatmann, T., Volk, M.: The probabilistic model checker Storm. Int. J. Softw. Tools Technol. Transf. **24**(4), 589–610 (2022)
13. Horváth, A., Telek, M.: Approximating heavy tailed behavior with phase-type distributions. In: Proceedings of 3rd International Conference on Matrix-Analytic Methods in Stochastic Models. Leuven, Belgium (2000)
14. Inc., W.R.: Mathematica, Version 14.0, Champaign, IL (2024). https://www.wolfram.com/mathematica
15. Kleinrock, L.: Queueing Systems, Vol. 1: Theory. Wiley (1975)
16. Kwiatkowska, M., Norman, G., Parker, D.: PRISM 4.0: verification of probabilistic real-time systems. In: Gopalakrishnan, G., Qadeer, S. (eds.) CAV 2011. LNCS, vol. 6806, pp. 585–591. Springer, Heidelberg (2011). https://doi.org/10.1007/978-3-642-22110-1_47
17. Marsan, M.A., Conte, G., Balbo, G.: A class of generalized stochastic Petri nets for the performance evaluation of multiprocessor systems. ACM Trans. Comput. Syst. **2**(2), 93–122 (1984)
18. Neuts, M.: Probability distributions of phase type. In: Liber Amicorum Prof. Emeritus H. Florin, pp. 173–206. University of Louvain (1975)
19. Ross, S.M.: On the time to first failure in multicomponent exponential reliability systems. Stochastic Process. Appl. **4**(2), 167–173 (1976)

20. Stewart, W.J.: Probability, Markov Chains, Queues, and Simulation: The Mathematical Basis of Performance Modeling. Princeton University Press (2009)
21. Wasserman, L.: All of Statistics. STS, Springer, New York (2004). https://doi.org/10.1007/978-0-387-21736-9

Under the Space Threat: Quantitative Analysis of Cosmos Blockchain

Daria Smuseva[1], Ivan Malakhov[2](✉), Andrea Marin[2], Carla Piazza[1], and Sabina Rossi[2]

[1] Università degli Studi di Udine, Udine, Italy
daria.smuseva@unive.it, carla.piazza@uniud.it
[2] Università Ca' Foscari Venezia, Venezia, Italy
{ivan.malakhov,marin,sabina.rossi}@unive.it

Abstract. Many contemporary blockchains are greatly dependent on the Proof-of-Stake (PoS) consensus protocol. Among these, Cosmos is a fairly new blockchain system that stands out as a prominent PoS example thanks to its ecosystem designed to facilitate interoperability between different blockchains through the Inter-Blockchain Communication protocol. Cosmos consensus protocol is called *CosmosBFT* and is based on the definition of rounds for consensus on blocks. The agreement has to be made by special users, namely validators, chosen among the participants with the highest bonded stakes. This paper investigates the potential new attacks that these design features can introduce. First, we show that the current state of nearly every network in the Cosmos ecosystem is prone to the unbalanced distribution of voting power (VP) of validators, as it is skewed towards a small group of already top-ranked ones. Secondly, we introduce a base model reflecting the standard execution of the consensus protocol of the Cosmos ecosystem. Then, we propose two case studies to assess the effect of network performance due to (i) colluded behavior of the highest-ranked validators and (ii) partial absence of the committee members. Our results suggest that a set of validators holding one-third of the total VP (also known as a *superminority*), either colluded or simultaneously unavailable, is able to drastically reduce the network effectiveness of producing blocks, thereby damaging honest network participants and network security.

Keywords: Blockchain · Proof-of-Stake · Performance Evaluation

1 Introduction

Blockchain technology, a decentralised and distributed peer-to-peer network that stores immutable data, has emerged as a transformative force across various sectors. Its ability to ensure transparency, security, and decentralisation has prompted widespread interest and adoption. Blockchains can be categorised into three primary groups: by access type, consensus mechanism, and smart contract functionality.

Firstly, blockchains can be classified by access type into public and private blockchains. Public blockchains are open to everyone, allowing unrestricted participation and fostering inclusivity and transparency. In contrast, private blockchains operate under restricted access policies, limiting participation to designated entities. The distributed nature of public blockchain systems introduces distinct security challenges, as the incentive structures often encourage participants to prioritise personal financial gain to maximise rewards.

Secondly, blockchains are classified by their consensus mechanisms, of which the two most popular are Proof of Work (PoW) and Proof of Stake (PoS). In PoW blockchains, such as the original Bitcoin blockchain introduced by Satoshi Nakamoto [14] in 2008, users called miners participate in the network by solving complex computational puzzles to obtain rewards. Miners receive rewards whenever they successfully create a block that is accepted by the network. This reward-driven participation model, however, brings several challenges, two of which have been discussed in our previous works [10–12,15]. In PoS systems, nodes do not rely heavily on computational resources to create blocks. Instead, consensus is achieved through a voting mechanism where the likelihood of becoming a block proposer depends on the amount of currency, or stake, a validator holds. For example, in Ethereum, the second largest blockchain after Bitcoin, participants must stake 32 ETH to become validators[1].

Lastly, blockchains can either support or do not support the smart contract functionality. The smart contracts are autonomous applications deployed on the blockchain ledger, consisting of code that facilitates and regulates interactions between parties without the need for intermediaries.

In this paper, we examine Cosmos[2] that is known to be a prominent example of PoS networks. Indeed, the Cosmos ecosystem is not merely a single blockchain network, but rather an entire ecosystem designed to facilitate interoperability among different blockchains within. Moreover, the choice of Cosmos stems from uniqueness of its consensus mechanism that is based on Tendermint protocol [7].

Validators play an important role in Cosmos networks, as well as in other Proof-of-Stake networks. They validate transactions and create blocks, thus maintaining the network's security and integrity. Validators stake their cryptocurrency as collateral, showcasing their dedication to the network's reliability. Unlike Ethereum, where each block is validated by a randomly chosen subset of an unlimited number of validators, Cosmos operates with a fixed number of validators. These validators are selected based on the highest stakes they hold.

The Cosmos ecosystem employs a common consensus mechanism that requires multiple steps for block commitment: Propose, Prevote, and Precommit. Each of these steps has default timeouts to manage the time required for block agreement. If validators fail to reach a consensus on a new block, they initiate a new round for the same block position until consensus is achieved. This approach ensures there are no forks but may result in lower network performance. A network experiencing multiple rounds for block approval would have reduced

---

[1] At the time of writing, this corresponds approximately to $3,300$ USD.
[2] https://cosmos.network.

throughput since transactions in these blocks would face delays in acceptance due to the protocol's requirements.

The Cosmos ecosystem presents an intriguing case study for exploring the dynamics of PoS blockchains and the existing challenges. One of them is Verifier's Dilemma which is based on the issue where the verification process lacks incentivisation. This concept assumes that some validators can accept new blocks without proper verification, undermining the integrity of the system [6]. We have already studied the phenomenon in Ethereum PoW and PoS [15,16], where there was a clear interest for users to behave unfair since they increase their reward. The concept of Tendermint blockchains does not imply voting incentives making the network even more exposed to the Verifier's Dilemma. However, the nature of Tendermint makes it difficult to evaluate the effect of the dilemma. The only way is to see how delays in voting steps can affect the network throughput.

Another issue within the Cosmos ecosystem is the voting power distribution. One-third of voting power is concentrated in the hands of a few validators. This concentration poses a risk to the network's integrity, as these validators, whether through collusion or accidental absence, can significantly impact the network.

In this paper, we study how such validators affect the network. We examine the implications of voting power concentration and analyse the potential risks associated with validator behavior. Our findings contribute to a deeper understanding of the dynamics within the Cosmos ecosystem and offer recommendations for improving its governance model.

Contribution. In this paper, (i) we provide a base analytical model that reflects the validation process using the Performance Evaluation Process Algebra (PEPA) tool presented in [8]. (ii) We introduce a scenario that extend the base model and study the case of colluded validators holding $\frac{1}{3}$ of the network voting power to discard proposed blocks of the target group. Finally, (iii) we describe another scenario which allows us to examine the case of partially absent validators that would be eventually unavailable in the network.

Paper structure. The paper is structured, as follows: Sect. 2 delves into Cosmos blockchain. Next, Sect. 3 introduces a set of analytical models based on the PEPA tool and covers assessment of the models. The experimental part is also included in this section. Finally, Sect. 4 concludes the paper and builds a basement for future work.

2 Background on Cosmos Blockchain

In this section, we provide a brief description of Cosmos network delving into the protocol underlying it.

Generally speaking, Cosmos serves as a platform enabling users to establish their own blockchains within its ecosystem. These blockchains, interconnected through Inter-blockchain Communication (IBC), maintain autonomy while facilitating communication. Operating on the CometBFT protocol, which is fully integral to Tendermint protocol, they retain flexibility for customization, although many projects adopt the default protocol settings. Additionally, the Cosmos platform hosts the Cosmos Hub, an early blockchain implementation often referenced as a model of their network protocol. In this context, we refer to Cosmos as a network instance rather than solely a platform.

In the network, anyone can attempt to become a validator, provided they stake a substantial amount of tokens to compete with the largest stakeholders. For instance, in Cosmos blockchain, the top five validators hold together more than 400 millions USD in fiat equivalent. An aspiring validator can achieve this stake not only by using their own tokens but by attracting users not involved in validation to delegate their funds to the validator in exchange for passive rewards (these users are known as *delegators*). It is known that the majority of the validators including top-ranked ones bond just bare minimum of tokens themselves, while the rest is aggregated thanks to the attracted delegators. Although validators can bond their own assets through delegation from accounts they control, this highlights the crucial role that delegators play within the network.

Network Consensus. The consensus process of CometBFT consists of several steps [1] that can be visualized as follows:

$$\texttt{NewHeight} \to (\texttt{Propose} \to \texttt{Prevote} \to \texttt{Precommit})^{\geq 1} \to \texttt{Commit} \to \ldots$$

In turn, three special steps, namely `Propose`, `Prevote`, and `Precommit` form a single Round. The round steps are repeated until more than $\frac{2}{3}$ of the total voting power precommits for the same block, which is then committed and added to the blockchain. If this does not occur within a timeout period, the next round is started with a new proposer but for the same height. Any PoS protocol including CometBFT ensures that the validators can reach consensus on a unique block as long as up to $\frac{1}{3}$ of the voting power is controlled by malicious or faulty validators.

Regarding the remaining steps, `NewHeight` serves for incrementing the height and assuring that most of the participants execute a commit step by waiting. Next, `Commit` is needed to perform a commit and move again to `NewHeight` step repeating the validation process.

Incentivisation. In Cosmos ecosystem, validators are mainly incentivised with block and transactions' fee rewards lacking any incentivisation for actual validation of blocks. Recall that in Cosmos all active validators bond their funds to form a stake. The reward that is accumulated during a certain reward period (mainly for each new block consolidation but also transaction fees from the accepted blocks of the period) is then distributed strictly proportional to the fraction of the stake the validators hold. Consequently, some validators who are not interested in full commitment to the validation process may decide to save their resources by not performing any of the steps described above without essentially any drawback. Moreover, considering the existing downtime penalty of 0.01% that implies missing more than 95% of the last 10,000 blocks which is approximately 19 hours of absence, malicious validators may decide to stay offline for exactly $10,000 * 0.95 - 1 = 9499$ blocks saving the real-world resources and still stay fully paid as the received reward is independent of the time spent on performing the validation process [3].

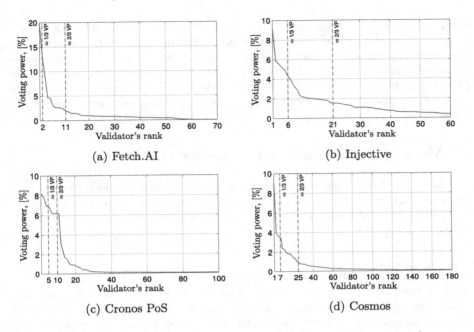

Fig. 1. Distribution of VP among all validators in the chosen networks.

There are few more ways to get reward as a validator. First, validators usually set a commission rate for all the reward received from using extra assets that were given by the delegators. The commission can have arbitrary value, however, it is quite common to see the rate of 10%. Although some committee members may retain all rewards from delegators with 100% commission (e.g., Kraken [5]). Thus, it is highly favorable for validators to attract more delegators.

Second, each block proposer has an opportunity to receive 5% of extra fee reward by including all the votes for the previous block into the one that they are entitled to propose. In simple terms, including just above two-thirds of the validators' votes the proposer would receive only 95% of the total fee in their block. The collection of all give them 4% of additional reward. Note that such reward is linked to the risk for a proposer. If they would wait too long to collect all the votes a proposer can simply miss their time and lose all the fee reward [3].

Voting Power Distribution. Figure 1 shows the voting power distribution among the leading blockchains within Cosmos ecosystem starting from Fetch.AI[3]. Clearly, among all the represented networks the supermajority, over two-thirds of network voting power, is held by a small group of top validators, leaving the rest of the network with minimal fractions of VP. Moreover, this concentration implies that if even a few members of this dominant groups become unavailable, as can be easily seen in Fetch.AI network where only *two* validators hold

---

[3] https://fetch.ai.

a superminority of voting power (one-third of the total), the network cannot achieve consensus, resulting in its shutdown. Furthermore, it is not necessary that only top-ranked validators become unavailable but generally any combination of validators holding the third of total VP can interrupt the consensus process just being absent at the same time. We use this reasoning to investigate the performance and reliability of the system in the scenarios described in the following section.

3 Model Descriptions and Examination

In this section, we introduce and describe some PEPA [8,9,13] models that reflect the consensus processes and two scenarios of interest in Cosmos ecosystem. Next, in the experimental part, we study the outcomes of the models and provide our educated interpretation.

3.1 Base Model

We first present a PEPA model (*base model*) that formally describes the consensus protocol. In the following sub-sections, we show the refined models for further analysis. Table 1 shows the model of a Cosmos blockchain within Cosmos ecosystem by using the process algebra language.

Upon a successfully created block the new consensus process starts at a step *NewHeight*, triggering *Round* that performs *Propose*, consequently. If successful, it progresses to *Prevote* or *NilPrevote*, depending on the success parameter w_1 of the Propose step where $w_1 \in [0,1]$. From *Prevote*, it moves to *Precommit* or *Unsuccess*, that again linked to the success parameter $w_2 \in [0,1]$ this time for the Prevote step. In turn, *NilPrevote* never progresses to *NewHeight*, but to *Unsuccess* with the rate β that always starts a *Round* again. *Precommit* step either leads to *Commit* with a success probability $w_3 \in [0,1]$ or initiates a new *Round* with the complementary probability $1 - w_3$.

Table 1. Base PEPA model of consensus protocol in Cosmos ecosystem.

NewHeight	$\stackrel{def}{=}$	$(nh, n).Round$
Round	$\stackrel{def}{=}$	$(r, n).Propose$
Propose	$\stackrel{def}{=}$	$(p, w_1\gamma).Prevote + (p, (1-w_1)\gamma).NilPrevote$
Prevote	$\stackrel{def}{=}$	$(pv, w_2\beta).Precommit + (pv, (1-w_2)\beta).Unsuccess$
NilPrevote	$\stackrel{def}{=}$	$(npv, \beta).Unsuccess$
Unsuccess	$\stackrel{def}{=}$	$(pc, \delta).Round$
Precommit	$\stackrel{def}{=}$	$(pc, w_3\delta).Commit + (pc, (1-w_3)\delta).Round$
Commit	$\stackrel{def}{=}$	$(c_i, \eta).NewHeight$
where		$\gamma = max\left(\dfrac{1}{t_1}, \dfrac{1}{T_1}\right), \quad \beta = max\left(\dfrac{1}{t_2}, \dfrac{1}{T_2}\right), \quad \delta = \dfrac{1}{T_3}, \quad \eta = \dfrac{1}{T_4}.$

Regarding the probabilities of success for the Propose and Prevote steps, namely w_1 and w_2, we assume that they represent the probabilities that two independent exponential random variables X_1 and X_2 with means t_1 and t_2 are less than the corresponding timeouts T_1 and T_2. Consequently, we obtain:

$$w_1 = Pr[X_1 \leq T_1] = 1 - e^{-\frac{1}{t_1}T_1}, \quad w_2 = Pr[X_2 \leq T_2] = 1 - e^{-\frac{1}{t_2}T_2}. \quad (1)$$

Parameters t_1, t_2 are the ones of the exponential random variables modeling the actual processing time of the Propose and Prevote steps, while T_1, T_2 refer to the standard step timeouts specified in the protocol configuration of a given blockchain. More precisely, to model the system behaviour we make the following assumptions:

- *Duration of steps.* The durations of steps are exponentially distributed with an expected duration equal to the timeout parameter associated with that step. The probability of successfully terminating that step before the timeout is computed as described by Equations (1). The introduction of deterministic timeouts is known to be challenging to be modeled in a Markovian model and although it could be approximated with the method of phases, this complicates the analysis. In this context, this assumption has a relatively small impact, since the goal of the model is not that of an accurate prediction of some performance indices rather that of comparing different scenarios created by malicious behaviours of some validators.
- *Fixed steps' timeout.* We assume that all successive rounds possess the same timeout values as the first round of the block height.
- *Successful Precommit step.* Given the negligible impact of the success probability at the Precommit step (w_3), we simplify our models by fixing its value at $w_3 \rightarrow 1$. Consequently, our focus shifts from the speed of node vote propagation through the network to the block processing steps, specifically w_1 and w_2.

Such assumptions remain true also for the models in the following subsection. Note that for the sake of simplicity, we assume that the processing times of the consensus steps, and hence the step rates, are computed with respect to their corresponding timeouts. Table 2 shows the values of such parameters used in Cosmos (Hub) network. The timeouts are derived from the default CometBFT protocol configuration and can be also seen in numerous blockchain instances within Cosmos ecosystem. We can then safely choose these settings for our study.

3.2 Model with Colluded Superminority

In this experiment, we are interested in studying the scenario in which the smallest achievable group of validators that holds the superminority, i.e., the top-ranked validators, colludes to prevent network from committing blocks of certain validators. In this way, the malicious behavior of colluded validators reduces the attractiveness of the affected validators to their delegators due to a lower rate of successful block proposals. Consequently, the delegators are likely to distribute

Table 2. Default round step limits for a network in Cosmos ecosystem.

Step name	Timeout parameter	Duration
Propose	T_1	3s
Prevote	T_2	1s
Precommit	T_3	1s
Commit	T_4	1s

Table 3. PEPA model of consensus protocol with the colluded validators in Cosmos ecosystem.

$NewHeight$	$\stackrel{def}{=}$	$(nh, n).Round$
$Round$	$\stackrel{def}{=}$	$(r, d_C n).Propose_C + (r, d_T n).Propose_T + (r, d_R n).Propose_R$
$Propose_C$	$\stackrel{def}{=}$	$(p, w_1\gamma).Prevote_C + (p, (1-w_1)\gamma).NilPrevote$
$Prevote_C$	$\stackrel{def}{=}$	$(pv, w_2\beta).Precommit_C + (pv, (1-w_2)\beta).Unsuccess$
$Precommit_C$	$\stackrel{def}{=}$	$(pc, w_3\delta).Commit_C + (pc, (1-w_3)\delta).Round$
$Commit_C$	$\stackrel{def}{=}$	$(c_C, \eta).NewHeight$
$Propose_T$	$\stackrel{def}{=}$	$(p, w_1\gamma).Prevote_T + (p, (1-w_1)\gamma).NilPrevote$
$Prevote_T$	$\stackrel{def}{=}$	$(pv, \beta).Unsuccess$
$Propose_R$	$\stackrel{def}{=}$	$(p, w_1\gamma).Prevote_R + (p, (1-w_1)\gamma).NilPrevote$
$Prevote_R$	$\stackrel{def}{=}$	$(pv, w_2\beta).Precommit_R + (pv, (1-w_2)\beta).Unsuccess$
$Precommit_R$	$\stackrel{def}{=}$	$(pc, w_3\delta).Commit_R + (pc, (1-w_3)\delta).Round$
$Commit_R$	$\stackrel{def}{=}$	$(c, \eta).NewHeight$
$NilPrevote$	$\stackrel{def}{=}$	$(npv, \beta).Unsuccess$
$Unsuccess$	$\stackrel{def}{=}$	$(pc, \delta).Round$

where $\gamma = max\left(\dfrac{1}{t_1}, \dfrac{1}{T_1}\right)$, $\beta = max\left(\dfrac{1}{t_2}, \dfrac{1}{T_2}\right)$, $\delta = \dfrac{1}{T_3}$, $\eta = \dfrac{1}{T_4}$, and $d_C = \dfrac{1}{3}$, $d_T \in [0, \dfrac{2}{3}]$, $d_R = 1 - (d_C + d_T)$, with $d_R \geq 0$.

their assets to validators with higher ratings, such as to the colluded validators, thereby increasing the colluders' revenue from delegators' commissions.

To this aim, we extend the model from the Table 1 in order to model such validators' shift and assess the impact of the network performance and fairness.

In this context, we state that a network is fair if its validators receive an amount of reward equal to their VP. Thus, the corresponding *coefficient of fairness* can be defined as a relationship between the fraction of the throughput produced by given validator(s) and the fraction of their possessed voting power. In general, the coefficient of fairness can be written as follows:

$$\phi_v = \frac{\frac{X_v}{X}}{\frac{VP_v}{VP}},$$

where $\frac{X_v}{X}$ refers to a fraction of throughput produced by a validator or set of validators v, while $\frac{VP_v}{VP}$ stands for fraction of VP possessed by v. For instance, the coefficient of fairness $\phi_v = 1$ would refer to the ideal case or the fair share of expended work among validators, while the values greater than 1 such that $\phi_v > 1$ would indicate extra efforts that such validator(s) have to put to maintain the network growth. Note that the coefficient of fairness greater than 1 can be a useful indicator that is capable of revealing such anomalies in the network and help with finding the source of the problem.

In the model, we introduce three new categories of validators, namely:

- *Attackers.* Attacking actors are the group of colluded validators possessing the superminority of VP. Their aim is to neglect the blocks of their target group by omitting the voting steps or voting for *nil*. Further, since they control the superminority of voting power, the network will fail to find consensus on the target blocks and would be required to move to new rounds.
- *Target.* Target actors are fair validators who will struggle to see their blocks accepted due to malicious behaviour of the attackers. Every time any target validator would propose a block at some round r, the rest of the validators' set do not reach the supermajority vote to commit it leading to the next round $r+1$.
- *Rest.* Such validators act honestly and are not affected by the actions of the attackers despite the decreased network performance due to the unnecessary extra rounds required to approach a new block height as a consequence of the attack.

Table 3 illustrates the model with the colluded validators. Recall that when the system starts *Round* a chosen validator has to perform a block proposal. Since we now have various types of actors in the system with the probability $d_C = \frac{1}{3}$ one of the colluded validators becomes a block proposer starting a round with $Propose_C$. Otherwise, either one of the validators from the group under attack or remaining validators perform the proposal of a block with the probabilities d_T and d_R, respectively. Parameter d_T changes with respect to the interval $[0, \frac{2}{3}]$ that agrees with the limit such that probabilities of moving to any of the actors is $d_C + d_T + d_R = 1$. All the probabilities correspond to the possessed VP of the introduced actors. For better visualisation, the underlying derivation graph of the model is depicted in Fig. 2.

The goal of the colluded validators is to intentionally ignore blocks proposed by the target validators ($Propose_T$ component). Thus, even successful block proposals leading to $Prevote_T$ always end up at $Unsuccess$, i.e., validators do not reach consensus on this block. Further steps for all validators follow the description of the Base Model (see Table 1). The idea of the division of the validators into three groups is to make the model perform analysis comparing validators' throughput.

Figure 4a and 4c show performance of the networks as functions of the VP fraction possessed by the attacked validators. The former plot demonstrates the network with standard consensus times correlating with the default timeouts, i.e.,

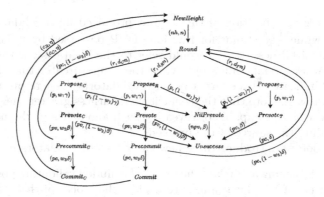

Fig. 2. Derivation graph of the model in Table 3 with colluded validators.

with average 6s per committed block like in Cosmos and many other blockchains within the ecosystem [2]. Such values of actual step times let us obtain the default probability for *Propose* and *Prevote* steps $w_{1,2} = 0.63$. Instead, the latter reflects the faster processing times than can be seen in Injective network with the average block time of 0.67s where $w_{1,2} \to 1$ that corresponds to $w_{1,2} = 0.99$ in our model [4]. Naturally, at the target VP fraction 0 (no attack is performed) the networks have the highest throughput of 0.08 and 1.5 blocks/s, respectively, that agrees with the intuition of the faster block times. Next, they gradually decrease up to a point when the colluded superminority vote only for blocks of themselves, thus the network throughput equals to the one of the attackers.

Figure 4b and 4d demonstrate the coefficients of fairness of the colluded validators, again, as functions of the VP fraction possessed by the affected validators. Initially, the plots show the ideal case when the networks are fair, so that $\phi_C = 1$ as there is no target for the attack. Later, we can see that the colluded validators start to notice the decrease of their effectiveness for both cases such that they have to perform more than they would normally do without running attack. When all honest validators become the target of the colluded ones the fairness of the latter group reaches the lowest point of one-third, i.e., they have to perform 3 times more to keep the network functionality, while drastically reducing throughput. Note that the plots for two configurations are identical, this is due to the fact that the proportional changes remain fixed for all variants. Thus, we can state that Fig. 4b (or 4d) is universal representation of the attack in terms of coefficient of fairness and fraction of target VP.

3.3 Model with Partially Absent Superminority

In the experiment, we focus on the scenario in which at each moment of time there is an arbitrary set of validators with the aggregated VP of one-third that are absent, i.e., they cannot or simply do not want to propose blocks themselves as well as verify blocks of other validators. Moreover, the validators' set is, in general, not fixed and can consist of different validators over time.

Table 4. PEPA model of consensus protocol with absent validators in Cosmos ecosystem.

$NewHeight \stackrel{def}{=}$	$(nh, n).Round$
$Round \stackrel{def}{=}$	$(r, d_A n).Propose_A + (r, (1 - d_A)n).Propose_R$
$Propose_A \stackrel{def}{=}$	$(p, aw_1\gamma).Prevote_A + (p, (1 - aw_1)\gamma).NilPrevote$
$Prevote_A \stackrel{def}{=}$	$(pv, w_2\beta).Precommit_A + (pv, (1 - w_2)\beta).Unsuccess$
$Precommit_A \stackrel{def}{=}$	$(pc, w_3\delta).Commit_A + (pc, (1 - w_3)\delta).Round$
$Commit_A \stackrel{def}{=}$	$(c_A, \eta).NewHeight$
$Propose_R \stackrel{def}{=}$	$(p, w_1\gamma).Prevote_R + (p, (1 - w_1)\gamma).NilPrevote$
$Prevote_R \stackrel{def}{=}$	$(pv, aw_2\beta).Precommit_R + (pv, (1 - aw_2)\beta).Unsuccess$
$Precommit_R \stackrel{def}{=}$	$(pc, w_3\delta).Commit_R + (pc, (1 - w_3)\delta).Round$
$Commit_R \stackrel{def}{=}$	$(c, \eta).NewHeight$
$NilPrevote \stackrel{def}{=}$	$(npv, \beta).Unsuccess$
$Unsuccess \stackrel{def}{=}$	$(pc, \delta).Round$

where $\gamma = max\left(\dfrac{1}{t_1}, \dfrac{1}{T_1}\right)$, $\beta = max\left(\dfrac{1}{t_2}, \dfrac{1}{T_2}\right)$, $\delta = \dfrac{1}{T_3}$, $\eta = \dfrac{1}{T_4}$, $d_A = \dfrac{1}{3}$, and $a \in [0, 1]$.

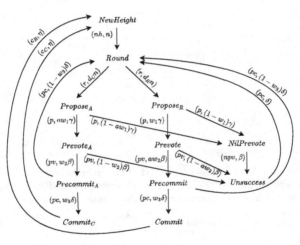

Fig. 3. Derivation graph of the model in Table 4 with partially absent validators.

We utilise the model from the Table 1 to reflect the absent validators in the new model of Table 4. It follows the similar logic of the model with colluded validators, however it has some differences. First, in this model there are only two types of actors, namely *absent* validators and the *rest*. Like in the previous model, they have their own probability parameters to start a round with

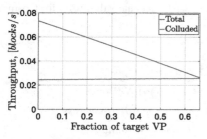
(a) Network throughput as a function of fraction of target validators' VP.

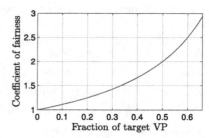
(b) Coefficient of fairness as a function of fraction of target validators' VP.

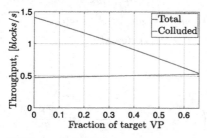
(c) Network throughput as a function of fraction of target validators' VP.

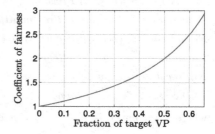
(d) Coefficient of fairness as a function of fraction of target validators' VP.

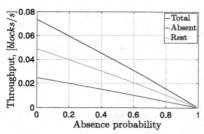
(e) Network throughput as a function of absence probability. $w_{1,2} = 0.63$, Cosmos ecosystem default configuration.

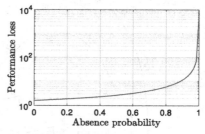
(f) Performance loss as a function of absence probability. $w_{1,2} = 0.63$, Cosmos ecosystem default configuration.

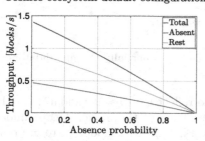
(g) Network throughput as a function of absence probability. $w_{1,2} = 0.99$.

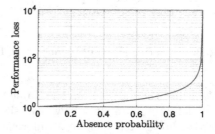
(h) Performance loss as a function of absence probability. $w_{1,2} = 0.99$.

Fig. 4. Performance evaluation of two network scenarios using PEPA model.

Propose, namely $d_A = \frac{1}{3}$ (the superminority) and $d_R = 1 - d_A$, respectively. Next, we redefine the probabilities of $Propose_A$ and $Prevote_R$ steps of absent validators and the rest, respectively, such that the likelihood of $Propose_A$ to successfully proceed to $Prevote_A$ and the probability to collect more than two-thirds of VP at $Prevote_R$ directly connects to the *absence probability* of such validators. We define an absence probability as complementary parameter to coefficient a introduced in the model such that $\bar{a} = 1 - a$. For instance, if in the network 20% of time the validators holding the superminority VP are absent we say that the partially absent validators have the probability parameter $\bar{a} = 0.8$ of proposing their blocks in time. With the same probability they have chance to prevote for blocks of all other validators within the proposal timeout and transit to $Precommit_R$. Note that the success probabilities for Propose and Prevote steps (w_1 and w_2) stay untouched.

The remaining model components inherit their behaviour from the Base model. The corresponding derivation graph is presented in Fig. 3.

Figure 4e and 4g show the dynamics of the networks' throughput with respect to the absence probability of the validators holding the superminority of VP. Clearly, for both configurations a descending pattern reassembles and at 50% chance of inactivity ($\bar{a} = 0.5$) we observe the half drop of the performance where the always absent validators cause complete stale of the network.

Figure 4f and 4h demonstrate the correlations between the performance inefficiency and the absence probability of the superminority validators. We define the *loss coefficient* as a proportion of the throughput of all proposed blocks and the blocks that were actually committed to the network. Using logarithmic scale for y-axis, we observe that both figures tend to exponentially grow with the increase of absence probability. Furthermore, at the absence probability 0 Fig. 4h shows non-zero performance loss due to probabilities w_1 and w_2 that are lower 1. Note that the figures can be also used to tell how many blocks the network has to offer to commit a single one of them. The higher the ratio, the more ineffective the network would be.

4 Conclusion

In this paper, we have explored the vulnerabilities inherent in the Proof-of-Stake consensus protocol within the Cosmos blockchain ecosystem. Our investigation centered on the CosmosBFT consensus protocol, which aims to provide fast finality and seamless interoperability among diverse blockchains via the Inter-Blockchain Communication protocol. Despite these advanced features, we identified significant security concerns related to the distribution of voting power among validators.

Our analysis demonstrates that the Cosmos network is susceptible to centralization risks where a small group of top-ranked validators holds a disproportionate amount of voting power. This unbalanced distribution creates potential vectors for attacks that could severely disrupt the network's functionality. Specifically, our quantitative case studies based on PEPA modelling technique revealed that a superminority, comprising validators with one-third of the total voting power, can significantly impair the network's block production if they engage in malicious behavior. Moreover, the latter scenario reveals that the network can tolerate the relatively low probability of absence of the superminority validators, while approaching 100% the network suffers from lack of the block throughput.

These findings highlight the need for further research on mitigating risks of validator centralization in PoS networks. Future work will explore approaches like Ethereum's random sub-committees to strengthen consensus protocols and preserve blockchain security and decentralization against emerging threats.

Acknowledgements. This study was carried out within the PE0000014 - Security and Rights in the CyberSpace (SERICS) and received funding from the European Union Next-GenerationEU - National Recovery and Resilience Plan (NRRP) - MISSION 4 COMPONENT 2, INVESTIMENT 1.3 - CUP N. H73C22000890001. This work has been also partially supported by the Research Project INDAM GNCS 2024 - CUP E53C23001670001 "Modelli composizionali per l'analisi di sistemi reversivili distribuiti (MARVEL)" and by the Project PRIN 2020 - CUP N. 20202FCJMH "NiRvAna - Noninterference and Reversibility Analysis in Private Blockchains". This manuscript reflects only the authors' views and opinions, neither the European Union nor the European Commission can be considered responsible for them.

References

1. CometBFT: consensus algorithm. https://docs.cometbft.com/main/spec/consensus/consensus. Accessed 15 Oct 2024
2. Cosmos blockchain dashboard on official cosmos explorer. https://www.mintscan.io/cosmos/. Accessed 15 Oct 2024
3. Cosmos network: validator incentives. https://hub.cosmos.network/validators/validator-faq.html#incentives. Accessed 15 Oct 2024
4. Injective blockchain dashboard on official cosmos explorer. https://www.mintscan.io/injective/. Accessed 15 Oct 2024
5. Kraken validator in Cosmos blockchain. https://www.mintscan.io/cosmos/validators/cosmosvaloper1z8zjv3lntpwxua0rtpvgrcwl0nm0tltgpgs6l7. Accessed 15 Oct 2024
6. Alharby, M., Lunardi, R.C., Aldweesh, A., Van Moorsel, A.: Data-driven model-based analysis of the Ethereum verifier's dilemma. In: 2020 50th Annual IEEE/IFIP International Conference on Dependable Systems and Networks (DSN), pp. 209–220. IEEE (2020)
7. Buchman, E., Kwon, J., Milosevic, Z.: The latest gossip on BFT consensus. arXiv preprint arXiv:1807.04938 (2018)
8. Hillston, J.: A Compositional Approach to Performance Modelling. Cambridge University Press (1996)
9. Hillston, J., Marin, A., Piazza, C., Rossi, S.: Persistent stochastic non-interference. Fund. Inform. **181**(1), 1–35 (2021)
10. Malakhov, I., Marin, A., Rossi, S.: Analysis of the confirmation time in proof-of-work blockchains. Future Gener. Comput. Syst. **147**, 275–291 (2023)

11. Malakhov, I., Marin, A., Rossi, S., Menasché, D.S.: Confirmed or dropped? Reliability analysis of transactions in PoW blockchains. IEEE Trans. Netw. Sci. Eng. **11**(4), 3276–3288 (2024)
12. Malakhov, I., Marin, A., Rossi, S., Smuseva, D.: On the use of proof-of-work in permissioned blockchains: security and fairness. IEEE Access **10**, 1305–1316 (2022)
13. Marin, A., Piazza, C., Rossi, S.: Proportional lumpability. In: André, É., Stoelinga, M. (eds.) FORMATS 2019. LNCS, vol. 11750, pp. 265–281. Springer, Cham (2019). https://doi.org/10.1007/978-3-030-29662-9_16
14. Nakamoto, S.: Bitcoin: A peer-to-peer electronic cash system (2008). http://www.bitcoin.org/bitcoin.pdf
15. Smuseva, D., Malakhov, I., Marin, A., van Moorsel, A., Rossi, S.: Verifier's dilemma in Ethereum blockchain: a quantitative analysis. In: International Conference on Quantitative Evaluation of Systems, pp. 317–336. Springer (2022). https://doi.org/10.1007/978-3-031-16336-4_16.pdf
16. Smuseva, D., Malakhov, I., Marin, A., Rossi, S.: Crisis of trust: analyzing the verifier's dilemma in Ethereum's proof-of-stake blockchain. In: 2023 IEEE International Conference on Blockchain (Blockchain), pp. 332–339. IEEE (2023)

A Lumped CTMC for Modular Rewritable PN

Lorentzo Capra[1]([✉]) and Marco Gribaudo[2]

[1] Dip. di Informatica, Università di Milano, Milan, Italy
capra@di.unimi.it
[2] Dip. di Elettronica, Informatica e Bioingeneria, Politecnico di Milano, Milan, Italy

Abstract. Petri Nets (PN) are extensively employed as a robust formalism for modelling concurrent and distributed systems, yet they struggle to model adaptive reconfigurable systems effectively. In response, we have developed a formalization for "rewritable" PT nets (RwPT) using Maude, a declarative language that upholds consistent rewriting logic semantics.

In this work we extend a recently introduced modular approach based on composite node labelling, to incorporate stochastic parameters, and we present an automated process to obtain a *lumped* CTMC from the quotient graph generated by a modular RwPT model. To demonstrate the efficacy of our method, we utilize a fault-tolerant manufacturing system as a case study.

Keywords: Maude · Reconfigurable systems · SPN · Lumped CTMC

1 Introduction

Despite their power, traditional formalisms such as Petri Nets, Automata, and Process Algebra do not offer designers a straightforward method to define dynamic system changes and assess their impact on performance and reliability. In most of cases, a modeler must explicitly define all the possible configurations with different sub-models, and add some logic to switch among them when needed. However, many fault-tolerant systems are equipped with self-reconfiguration capabilities to ensure the desired quality of service. Therefore, various extensions to these classical models have been made. have been proposed. Rewritable PT nets (RwPT) were presented in [8] as a versatile formalism for the modeling and analysis of adaptive distributed systems. The steps of RwPT were defined using the Maude declarative language, which employs Rewriting Logic to provide both operational and mathematical semantics, thus creating a scalable model for self-adapting PT nets. Unlike similar approaches [16], which translate a simpler type of PNs into Maude, the RwPT formalism streamlines data abstraction, is concise and efficient, and circumvents the limitations imposed by the use of push-out typical in graph transformation systems. Other approaches include, for instance, exception mechanisms, which are very complex to handle [2].

RwPT is an extension of Graph Transform Systems (GTS). Considering graph isomorphism (GI) when identifying equivalent states within the model dynamics is essential: this feature is highly advantageous for scaling up the model's size or parallelism degree, especially when integrating a Stochastic Process into the model's state space. Recent studies have shown that GI has a quasi-polynomial complexity [1]. Graph Canonization (GC) $can(G)$, which is at least as complex as GI, involves finding a canonical form for any graph such that for any two graphs G and G', if $G \simeq G'$, then $can(G) = can(G')$.

We developed a general canonization method [7] to be used with RwPT, which is integrated into Maude. In [9], the approach is extended for constructing extensive RwPT models using algebraic operators. By employing composite node labeling, we detect symmetries and preserve a hierarchical structure through net rewrites. By merely permuting labels, we achieve a normal form for PT net terms. The ideas proposed here are not limited to PT net, and can be used in generic frameworks, supporting multi-formalism modeling, such as [14] and [15].

In this paper, we utilize a case study, taken from the literature, to introduce an automated method for constructing a Lumped Continuous-Time Markov Chain (CTMC) from the quotient graph of an RwPT model, after incorporating stochastic parameters into the system. Section 2 provides background information, which is followed by a detailed example in Sect. 3. The modular RwPT formalism, now enhanced with stochastic parameters, is discussed in Sect. 4. In Sect. 5, we describe the procedure to derive a lumped CTMC from an RwPT model and validate its effectiveness with experimental results on classical dependability metrics. We conclude with a discussion of ongoing research.

2 (Stochastic) PT Nets and Maude

The implementation of the PT formalsim is based on multisets. For a given set D, a *multiset* (or *bag*) b over D is a function $b : D \to \mathbb{N}$, where $b(d)$ represents the *multiplicity* of an element d in b. The collection of all multisets on D is denoted as $Bag[D]$. A stochastic PT (or SPN) *net* [12,17] is a 6-tuple (P, T, I, O, H, λ) representing a bipartite graph, where: P, T are finite, non-empty, disjoint sets holding places and transitions; the functions $\{I, O, H\} : T \to Bag[P]$ describe the *input*, *output*, and *inhibitor* edges as transition incidence matrices; $\lambda : T \to \mathbb{R}^+$ assigns each transition a negative exponential firing rate.

A PT net *marking* (state) is a multiset $m \in Bag[P]$. The PT net dynamics is defined by *firing rule*: $t \in T$ is *enabled* in the marking m if and only if:

$$I(t) \leq m \wedge \forall p \in P \; H(t)(p) = 0 \vee H(t)(p) > m(p)$$

If t is enabled in m it may fire, leading to marking m' (we denote this $m[t\rangle m'$)

$$m' = m + O(t) - I(t)$$

A PT-*system* is a pair (N, m_0), where N is a PT net and m_0 is a marking of N. The interleaving semantics of (N, m_0) is specified by the *reachability graph*

(RG), an edge-labelled, directed graph (V, E) whose nodes are markings. It is defined inductively: $m_0 \in V$; if $m \in V$ and $m[t\rangle m'$ then $m' \in V$, $m \xrightarrow{t} m' \in E$.

The timed semantics of a stochastic PT system is a CTMC isomorphic to the RG. For any two $m_i, m_j \in V$, the transition rate from m_i to m_j is $r_{i,j} := \sum_{t:m_i[t\rangle m_j} \lambda(t)$. The CTMC infinitesimal generator is a $|V| \times |V|$ matrix Q such that $Q[i,j] = r_{i,j}$ if $i \neq j$, $Q[i,i] = 1 - \sum_{j,j\neq i} r_{i,j}$.

The Maude system Maude [13] is a highly expressive, purely declarative language characterized by a rewriting logic semantics [4]. Statements consist of (conditional) *equations* and *rules*. Each side of a rule or equation represents terms of a specific *kind* that might include variables. The semantics of rules and equations involve straightforward rewriting, where instances of the left-hand side are substituted by corresponding instances of the right-hand side. The expressivity of Maude is realized through the use of matching modulo operator equational attributes, sub-typing, partiality, generic types, and reflection. A Maude *functional* module comprises only *equations* and functions as a functional program defining one or more operations through equations, utilized as simplification rules. A functional module details an *equational theory* within membership equational logic [3]. Formally, such a theory is a tuple $(\Sigma, E \cup A)$, with Σ representing the signature, which includes the declaration of all sorts, subsorts, kinds[1], and operators; E being the set of equations and membership axioms; and A as the set of operator equational attributes (e.g., assoc). The model of $(\Sigma, E \cup A)$ is the *initial algebra* $T_{\Sigma/E \cup A}$, which mathematically corresponds to the quotient of the ground-term algebra T_Σ. Provided that E and A satisfy nonrestrictive conditions, the final (or *canonical*) values of ground terms form an algebra isomorphic to the initial algebra, ensuring that the mathematical and rewriting semantics are identical.

A Maude *system module* includes *rewrite rules* and, potentially, equations. These rules illustrate local transitions in a concurrent system. In formal language, a system module outlines a generalized *rewrite theory* [4], symbolized as a four-tuple $\mathcal{R} = (\Sigma, E \cup A, \phi, R)$, where $(\Sigma, E \cup A)$ constitutes a membership equational theory; ϕ identifies the frozen arguments for each operator in Σ; and R contains a set of rewrite rules[2]. This rewrite theory models a concurrent system. $(\Sigma, E \cup A)$ establishes the algebraic structure of the states, while R and ϕ define the concurrent transitions of the system. The initial model of $\mathcal{R}$ assigns to each kind k a labeled transition system (TS) where the states are the elements of $T_{\Sigma/E \cup A,k}$, and the state transitions occur as $[t] \xrightarrow{[\alpha]} [t']$, with $[\alpha]$ representing a class of rewrites of *equivalent*. The property of *coherence* guarantees that a strategy that reduces terms to their canonical forms before applying the rules is sound and complete. A Maude system module is also an executable specification of distributed systems.

---

[1] Kinds are implicit equivalence classes defined by connected components of sorts (as per subsort partial order). Terms in a kind without a specific sort are *error* terms.

[2] Rewrite rules do not apply to frozen arguments.

3 Running Example: A Fault-Tolerant MS

The illustrative example in this paper depicts a distributed production system that gracefully degrades, shown by the two PT systems in Fig. 1. Although it is a literature example, it has been used as a reference [6,9] due to its challenging adaptation issues involving both the structure and the state of a system.

The left net represents a Production Line (denoted PL) which is divided into K lines (robots) that handle raw pieces (a multiple of K). These branches ($\{w_i, ln_i, a_i\}$, $i : 0 \ldots K-1$) are fully interchangeable. An assembly component (transition as) converts the processed pieces K into an artifact. A loader (ld) collects K pieces from a storage facility (place s) on the lines. In this study, $K = 2$. The initial count of pieces (tokens) in s is $K \cdot M$, where $M \in \mathbb{N}^+$ is another parameter of the model. For each artifact produced, K new pieces are introduced. A branch might fail (transitions ft_i). When that occurs, the PL restructures to continue functioning, but with reduced capacity. Simple static analysis can show that the PL system reaches a *deadlock* after a failure.

The net on the right of Fig. 1 shows the development of PL after a fault occurs (considering scenario $K = 2$). This process involves moving pieces from the faulted branch to the remaining branch(es) to maintain the production cycle. Traditional PN frameworks (including high-level PN variants) cannot model this operation. The items left on the faulty line (represented as place w_1 here) are transferred to the remaining functional line (w_0): The marking of the PT net at the bottom demonstrates the state after adaptation. We assume that a PL that fails twice is beyond repair.

We will examine a situation in which N PL replicas function simultaneously and degrade gracefully, as shown in Fig. 2. This behavior can be extended to a PT system that incorporates N PLs, each operating K parallel lines that handle $K \cdot M$ raw items, denoted by the term NPLsys(N, K, M). The development proceeds in two phases: *i)* when a fault impacts a PL, it autonomously adjusts to continue functioning in a reduced capacity, *ii)* when a fault occurs in a degraded PL, the entire system disconnects it (the final step in Fig. 2), and the remaining items are then relocated to the warehouse.

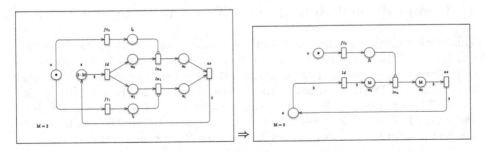

Fig. 1. Production Line (PL) and adaptation after a fault.

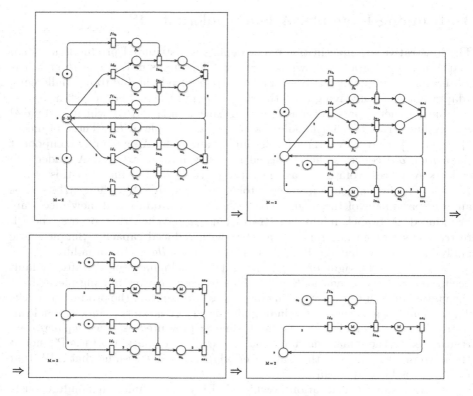

Fig. 2. A possible path of the Gracefully Degrading Production System.

Describing the system's structural changes and the concurrent transfer of items from a malfunctioning line to a functioning one using classical PN formalisms, even in their high-level versions, poses a challenging task for expert modellers.

4 Compositional Rewritable PT and their Symmetries

This section introduces the concept of rewritable stochastic PT nets (RwSPT), which extend the *modular* rewritable PT nets described in [9] by incorporating transitions with stochastic parameters. An RwSPT serves as an algebraic model of a stochastic PN [12], combining the rewrite of rules with the PT firing mechanism. The definition of RwSPT includes a hierarchy of Maude modules (e.g., BAG, PT-NET, PT-SYSTEM) most of which are described in [9]. The Maude sources can be found in https://github.com/lgcapra/rewpt/tree/main/modSPT.

The RwSPT definition uses structured annotations to underline the symmetry of the model. It features a concise place-based encoding to aid in state canonization and is based on the functional module BAG{X}, which introduces multisets as a complex data type. Specifically, the commutative/associative _+_ operator provides an intuitive way to describe a multiset as a weighted sum. For example, 3 . a + 1 . b is a multiset with 3 instances of a and one of b.

The sort Pbag contains multisets of places. Each place label (a term of sort Plab) is a nonempty list of pairs built of String and a Nat. Places are uniquely identified by their labels. These pairs represent a symmetric component within a nested hierarchy. Compositional operators annotate places incrementally from right to left, with the label suffix representing the root of a hierarchy. For example, the 'assembly' place of line 1 in Production Line 2 would be encoded as: p(< "a"; 0 > < "L"; 1 >) .

We describe net transitions (terms Tran) through their incidence matrix (a triplet of terms Pbag) and associated tags. A tag includes a String, a Nat (indicating a priority), and a Float (interpreted as a firing rate).

$$[I,O,H] \; |\!\!-\!\!> \; <\!\!< S, P, R >\!\!>$$

When using the associative composition operator _;_ and the subsort relation Tran < Net, nets are defined in a modular way. For example, the subnet containing transitions ld and ln_0 in Fig. 1 (left) become the Net term in the Listing 1.1 (the zero-arity operator nilP represents an empty multi-set).

Listing 1.1. A (sub)net

```
[2 . p(< ''s'' ; 0 >), 1 . p(< ''w'' ; 0 >) + 1 . p(< ''w'' ; 1 >), nilP] |->
<< ''ld'', 0, 0.5 >> ;
[1 . p(< ''w'' ; 0 >), 1 . p(< ''a'' ; 0 >), 1 . p(< ''f'' ; 0 > ] |-> << ''ln'', 0, 0.1 >>
```

A System term is the empty juxtaposition (__) of a Net and a Pbag (representing the net marking). The conditional rewrite rule firing specifies the PT firing rule[3], as shown in the Listing 1.2.

Listing 1.2. PT Firing Rule

```
vars N N' : Net . var T : Tran . var M : Pbag .
crl [firing] : N M => N fire(T, M) if T ; N' := N /\ enabled(T, N M) .
```

The predicate enabled takes concession into account and is based on hasConcession, which determines the 'topological' aspect of enabling:

Listing 1.3. PT Firing operators

```
vars I O H M : Pbag . var L : Tlab .
op hasConcession : Tran Pbag -> Bool .
eq hasConcession([I,O,H] |-> L, M) = I <= M and-then H > B .
op fire : Tran Pbag -> Pbag .
eq fire([I,O,H] |-> L, M) = M + O - I .
```

---

[3] Notice the use of a matching equation: The free variables T, N', are matched (:=) against the canonical ground term bound to the variable N.

A RwSPT is defined by a system module that contains two constant operators, used as for aliasing: op net : -> Net and op m0 : -> Pbag.

Two equations define their bindings. This module includes rewriting rules R of System type incorporating firing. In this study, we use a fully non-deterministic approach (interleaving semantics). Rewrites are given equal priority and have an exponential rate (indicated in the rule label except for firing), such that for the state transition system, the following holds ($\subseteq$ denotes the subgraph relation): $TS(\text{net m0}, \{\text{firing}\}) \subseteq TS(\text{net m0}, R)$. Note that TS on the left corresponds to the standard RG.

We have provided net-algebra and net-rewriting operators [9] with a twofold intent: to ease the modeler's task and to enable the construction and modification of large-scale models with nested components by implicitly highlighting their symmetry. A TS *quotient* is built through a simple permutation of node labels. This method shows better performance compared to [7] (with similar state reduction) and is more effective than a well-known framework that utilizes symmetries, such as symmetric nets [10].

In a context where nets have a mutable structure, identifying behavioral equivalences reduces to a graph *morphism*. PT system morphism must maintain the edges and the marking: In our encoding, a *morphism* between PT systems (N m) and (N' m') is a bijection ϕ : places(N) $\to$ places(N') such that, considering the homomorphic extension of ϕ on multisets, $\phi(\text{N}) = \text{N'}$ and $\phi(\text{m}) = $ m'. Moreover, ϕ must retain the textual annotations of the place labels and the transition tags. If N' = N we speak of *automorphism*, in which case ϕ is a permutation in the set of places. We refer to a *normal* form that principally involves identifying sets of automorphic (permutable) places. Two markings m, m' of a net N are said automorphic, m $\cong$ m', if there is an automorphism ϕ in N that maps m into m'. The equivalence relation $\cong$ is a congruence, that is, it preserves the transition firings and *rates*.

Definition 1 (Symmetric Labeling). *A* Net *term is symmetrically labeled if any two maximal sets of places whose labels have the same suffix (possibly empty), which is preceded by pairs with the same tag, are permutable. A* System *term is symmetrically labeled if its* Net *subterm is.*

In other words, if a Net term N meets Definition 1, then for any two maximal subsets of places P, P', if L, L', L" : Plab, w: String, i, j : Nat :
$P := \{\text{p(L' < w ; i > L)}\}$, $P' := \{\text{p(L'' < w ; j > L)}\}$,
there exists an automorphism (permutation) ϕ such that $\phi(P) = P'$, $\phi(P') = P$, which is extended as an identity to the rest of places.

If a System term respects the previous definition, it can be transformed into a 'normal' form by merely swapping indices on the place labels (e.g., i $\leftrightarrow$ j), while still complying with Definition 1. This normal form is the most minimal according to a lexicographic order within the automorphism class ($\cong$) implicitly defined by Definition 1. However, in contrast to general graph canonization, there is no need for any pruning strategy or backtracking. In simple terms, a monotone procedure is used where the sequence of index swaps does not matter (see [9] for full details). Efficiency is achieved as the normalized form of the subterm

Net is derived through basic "name abstraction", where at each hierarchical level the indices of structured place labels continuously span from 0 to $k \in \mathbb{N}$.

The approach offers a streamlined set of operators that maintain the symmetric labeling of nets. This set features *compositional* operators and operators for *modifying* nets, such as adding or removing components. Rewriting rules necessitate these operators to handle System terms constructed in a modular fashion. Furthermore, the rules must comply with parametricity conditions (not detailed here) that constrain the use of non-variable terms. Rewriting rules that meet these conditions are referred to as *symmetric* [9].

Under these assumptions, we get a *quotient* TS from a System term that retains reachability and meets strong bisimulation. Let t, t', u, u' be (final) terms of sort System, r a System rule $r : s \Longrightarrow s'$. The notation $t \xRightarrow{r(\sigma)} t'$ (t is rewritten into t' through r) means that there is a ground substitution σ of r's variables such that $\sigma(s) = t$ and $\sigma(s') = t'$.

Property 1. Let t meet Definition 1 and r be a symmetric rule.

If $t \xRightarrow{r(\sigma)} t'$ then $\forall u, \phi, t \cong_\phi u: u \xRightarrow{r(\phi(\sigma))} u'$, $t' \cong u'$ (u', t' meet the Definition 1)

The TS quotient produced by a term $\hat{t}$ is achieved by applying the (overloaded) operator normalize to the right-hand side of the rewriting rules.
op normalize : System -> System . op normalize : Pbag -> Pbag .
When a System is rewritten using the firing rule, only the marking subterm obtained applying the operator normalize to fire(T, M) in Listing 1.2 is needed.

According to Property 1, if the morphism (index exchange) ϕ preserves the transition rates and the rules are parameterized, we can map the TS quotient of $\hat{t}$ onto a "lumped" CTMC. In a Markov process's state space, an equivalence relation is considered "strong lumpability" if the cumulative transition rates between any two states within a class to any other class remain consistent. Despite the possibility of establishing "exact lumpabability" [5], our attention is primarily focused on aggregated probability. Let us focus on system described in Sect. 3. It is composed of N Production Lines (PL) that share raw materials, with each PL split into K interchangeable lines (see Listing 1.4).

Listing 1.4. Modular Specification of a Fault Tolerant Production System

```
fmod FTPL is
 pr NET-OP{SPTlab} .
 ops PL PLA nomPL faultyPL NfaultyPL : NzNat -> Net .
 op faultySys : NzNat -> System .
 op NPL : NzNat NzNat -> Net [memo] .
 op NPLsys : NzNat NzNat NzNat -> System .
 ops loadLab asLab failLab workLab : -> Tlab [memo] .
 eq loadLab = << ''ld'',0, 0.5 >> .
 eq asLab = << ''as'',0, 2.0 >> .
 eq workLab = << ''ln'',0, 0.1 >> .
 eq failLab = << ''ft'',0, 0.001 >> .
 var I : Nat .
 vars N K M : NzNat .
 eq line = [1 . p(< ''w'' ; 0 >),1 . p(< ''a'' ; 0 >),1 . p(< ''f'' ; 0 >) ] |-> workLab .
 eq fault = [1 . p(< ''o'' ; 0 >) , 1 . p(< ''f'' ; 0 >), nilP ] |-> failLab .
```

```
eq load = [1 . p(< ''s'' ; 0 >) , 1 . p(< ''w'' ; 0 >) , nilP ] |-> loadLab .
eq ass  = [1 . p(< ''a'' ; 0 >) , 1 . p(< ''s'' ; 0 >) , nilP ] |-> asLab .
eq cycle = load ; line ; ass ; fault .
eq PL(K) = repl&share(cycle, K, ''L'', p (< ''o'' ; 0 >) U p(< ''s'' ; 0 >), asLab U
    loadLab) .
eq NPL(N, K) = repl&share(PL(K), N, ''PL'', p(< ''s'' ; 0 >), emptyStlab) .
eq NPLsys(N, K, M) = setMark(setMark(NPL(N, K), ''o'' ''PL'', 1), ''s'', K * M) .
endfm
```

We start by defining the net transitions. Then we build a Production Line using the `repl&share` operator: The term PL(K) represents a line with K symmetric branches, similar to the one shown in Fig. 1 (left). The structure of the submodel is expressed by adding a pair with the tag "L" to the place labels. For example, p(< "w" ; 0 > < "L" ; 1 >) describes the "working" place of line 1 of the Production Line. We can also choose to exclude places to share among replicas: In this case, we exclude those representing the "warehouse" (tag "s") and faults (tag "o"). Additionally, we can indicate transitions to share: For instance, "load" and "assembly" are shared.

The term NPL(N, K) of the type Net consists of N PLs, each of which contains K branches. This net was generated using the `repl&share` operator, which adds the "PL" tag to place labels to indicate an additional nesting level. The sharing mechanism ensures each PL gathers K raw pieces. The PT net represented by NPL(2,2) can be seen in Fig. 2, top-left (basic, unadorned labels are used in that representation). Furthermore, the term NPLsys(N, K, M) in System sort is a PT system that holds K*M tokens in the "warehouse" place, with a single token in each place tagged with "o" to trigger fault occurrences within a PL. We can build an identical model using the "symmetric" version of the process algebra ALT operator. The System term generated using the above operators possesses symmetrical labeling, and its Net subterm has already been normalized. Consider, e.g., NPLsys(2, 2, 1). When firing either of the conflicting "ld" transitions, the resulting markings are as follows:

m_1 p(< "o"; 0 > < "PL"; 0 >) + p(< "o"; 0 > < "PL"; 1 >) +
    p(< "w"; 0 > < "L"; 0 > < "PL"; 1 >) +
    p(< "w"; 0 > < "L"; 1 > < "PL"; 1 >)
m_2 p(< "o"; 0 > < "PL"; 0 >) + p(< "o"; 0 > < "PL"; 1 >) +
    p(< "w"; 0 > < "L"; 0 > < "PL"; 0 >) +
    p(< "w"; 0 > < "L"; 1 > < "PL"; 0 >).

These are automorphic (one can be converted into the other by interchanging < "PL"; 1 > ↔ < "PL"; 0 >), but the second marking is the smallest in lexicographic order and hence corresponds to the normalized form. The rewrite rule in Listing 1.5 encapsulates the self-adjustment of a PL with $K = 2$ in response to a fault, enabling it to function in a diminished capacity. This rule deviates slightly from [9], as it is locally activated by a breakdown, leading to a significantly larger TS. The rule only employs operators that uphold the Definition 1, such as `join` , `detach` , `setMark` , reinting the symmetrical labeling (Definition 1). A similar rule removes a faulty and degraded PL from the system.

Listing 1.5. Rewrite rule of a PL (the label contains the rule's exponential rate)

```
vars S S' S'' : Pbag . vars I J : Nat . vars Sys Sys' : System . var L : Lab .
crl [r1-0.005] : N S => normalize(join(Sys, setMark(setMark(
  Sys', ''w'' ''fPL'', | match(S', ''w'') |), ''a'' ''fPL'', | match(S', ''a'') |)))
  if S'' + 1 . p(< ''f'' ; J > L < ''PL'' ; I >) := S /\ N' := nomPL(I) /\ dead (N' S) /\
  S' := subag(S'', < ''PL'' ; I >) /\ Sys := detache(N, N') S'' - S' /\
  Sys' := faultySys(notIn(N,''fPL'')).
```

With the model-checking facilities of Maude (in this case, the search command), we can demonstrate that for any given N, the quotient transition system has two absorbing states: Every state comprises a deteriorated PL that contains all $2 \cdot M$ parts (unprocessed, except possibly one). This is equivalent to the command below, which yields the same results as its unconditioned counterpart.

```
search NPLsys(N,2,M) =>! F:System such that
net(F:System) == faultyPL /\ B:Pbag := marking(F:System)
/\ | match(B:Pbag, "w") | + | match(B:Pbag, "a") | == 2 * M
```

5 Getting the Lumped CTMC Generator from RwPT

The CTMC generator entry $Q[i,j]$ is defined as: $\sum_{r \in R} \lambda_r \cdot |S_{i,j}^r|$, where $\lambda_r \in \Re^+$ is a given rate, and $S_{i,j}^r = \{\sigma \mid \hat{t}_i \stackrel{r(\sigma)}{\Longrightarrow} t_j, t_j \cong \hat{t}_j\}$ represents the matches of r resulting in equivalent states. Therefore, to obtain the CTMC infinitesimal generator, it is necessary to quantify instances that correspond to a specific state transition. Our solution uses two operators: the first identifies potential matches for each rule based on the subset of independent variables involved, and the second simulates the rewriting process. These two operators can be "mechanically" defined from the syntax of a rule.

To gain a clearer understanding of the concept, let us examine a simplified scenario that encompasses the vast majority of cases and to which any case can be reduced. We suppose that for every rule $r \in R$: $i)\, r$ is "injective": if $t \stackrel{r(\sigma)}{\Longrightarrow} t' \wedge t \stackrel{r(\sigma')}{\Longrightarrow} t'$ then $\sigma = \sigma'$, $ii)$ if r is conditional ($r : s \Longrightarrow s'\ if\ cond$) the condition does not contain rewrite expressions ($u \Longrightarrow u'$). Based on these assumptions, we can automatically extend a stochastic RwPT specification to generate a redundant quotient TS that contains all the necessary information to construct the lumped CTMC generator.

The list 1.6, which is related to the running example, has a universal structure. To avoid overly technical details of Maude syntax, we outline an operator, rule, which encodes all rewriting rules except for firing (handled separately for efficiency). This operator defines a *partial* mapping where, given a label (defined using a Tlab) and a System term, it determines the corresponding term-rewriting if feasible: each rewrite rule is tied to an equation. The operator ruleApp builds upon rule: it computes all potential outcomes of rewriting that term using the rule. It does not execute term normalization. As is typical in Maude, the ruleApp definition is optimized via tail-recursion. Lastly, ruleExe, which extends ruleApp, partitions the results of a rule application to a term into "equivalence classes" (sort Rset) through normalization: each class is represented by a pair System <-| Float , that is the aggregate rate towards a normalized state. The operator ruleApp serves as the bulk form of ruleExe.

The excerpt in the Listing 1.7 illustrates the augmented state representation which contains detailed information on the normalized state transition. The state structure defined by the mixfix constructor StateTranSys comprises four fields.

Listing 1.6. rule encoding for the lumped CTMC

```
vars N N' N'' : Net . vars S S' S'' : Pbag . vars I Imin J : Nat .
vars Sys Sys' : System . var L : Lab . var Sp : Pset . var TL : Tlab .

op rule : Tlab System -> [System] . *** one equation for rule
ceq rule(<< ''r1'',0, 0.005 >>, N S) = join(Sys, setMark(setMark(Sys', ''w''
    fPL'', | match(S', ''w'') |), ''a'' ''fPL'', | match(S', ''a'') |))
  if S'' + 1 . p(< ''f'' ; J > L < ''PL'' ; I >) := S /\ N' := nomPL(I) /\
    dead (N' S) /\ S' := subag(S'', < ''PL'' ; I >) /\
    Sys := detache(N, N') S'' - S' /\ Sys' := faultySys(minNotIn(N, ''fPL'')) .
ceq rule(<< ''r2'',0, 0.01 >> , N S) = N'' set(S'' - S', p(< ''s'' ; 0 >),
    S[p(< ''s'' ; 0 >)] + | S' |) if S'' + 1 . p(< ''f'' ; J > L < ''fPL'' ; I >)
        := S
  /\ N' := faultyPL(I) /\ dead(N' S) /\ N'' := detache(N, N') /\
  N'' =/= emptyN /\ S' := subag(S'', < ''fPL'' ; I >) .

*** ''rule application'' (without normalization)
var SS : Set{System} . var TS : [System] . vars R F : Float .
ops ruleApp : Tlab System -> Set{System} .
eq ruleApp(TL, Sys) = $ruleApp(TL, Sys, emptySS) .
op $ruleApp : Tlab System Set{System} -> Set{System} .
ceq $ruleApp(TL, Sys, SS) = $ruleApp(TL, Sys, SS U TS) if TS := rule(TL, Sys)
    /\ TS :: System /\ not(TS in SS) .
 eq $ruleApp(TL, Sys, SS) = SS [owise] .
*** ''aggregate'' rates calculation (with normalization)
op rulexe : Tlab System -> Rset .
eq rulexe(TL, Sys) = $rulexe(rate(TL), ruleApp(TL, Sys), emptyRset) .
op $rulexe : Float Set{System} Rset -> Rset .
eq $rulexe(F, emptySS, RS) = RS .
ceq $rulexe(F, Sys U SS, RS ; Sys' <-| R) = $rulexe(F, SS, RS ; Sys' <-| R + F
    ) if Sys' := normalize(Sys) .
eq $rulexe(F, Sys U SS, RS) = $rulexe(F, SS, RS ; normalize(Sys) <-| F)
    [owise] .
op allRew : System -> Rset [memo] . *** bulk application
eq allRew(Sys) = rulexe(labr1, Sys) U rulexe(labr2, Sys) .
```

The initial pair describes the PT system, while the remaining two fields detail the state transitions caused by the firing rule and other rewrites, in that order. As explained, we collect state transitions (rule applications) that share the target for calculating aggregated rates. Let us consider the firing rule: it has two associated operators, namely enabSet, which gives the set of possible transitions, and fire, which provides the reached markings, each associated with the corresponding cumulative rate. The operator toStateTran converts the conventional state representation into a structured one which highlights aggregate transition rates. The implementation of the firing rule and other rewriting rules is straightforward, as their impact is directly reflected in the state information.

Listing 1.7. TS encoding for the lumped CTMC

```
vars B B' M M' : Pbag . var N : Net . var TS : TagSet . var FS : Fset . var RS : Rset .
var R : Float .
*** description of a system pointing out (aggregate) state-transition rates
op NET:_ M:_ FIRING:_REW:_ : Net Pbag Fset Rset -> StateTranSys [ctor] .
op toStateTran : System -> StateTranSys .
eq toStateTran(N M) = NET: N M: M FIRING: fire(enabSet(N M), M) REW: allRew(N M) .
*** caculates the cumulative firing effect of a net (set of transitions)
op fire : Net Pbag -> Fset .
*** definition of fire
*** firing rule
rl [firing] : NET: N M: B FIRING: (B' <-| R ; FS) REW: RS => toStateTranSPN(N B') .
*** structural rewrites
rl [rew] : NET: N M: B FIRING: FS REW: (Sys <-| R ; RS) => toStateTranSPN(Sys) .
```

When considering `toStateTran(NPLsys(2,2,2))`, which aligns with the PT net at the top of Fig. 2, the resulting quotient TS comprises 295 states compared to the 779 states in the standard TS. The quotient graph's state transitions often correspond to multiple matches. For instance, the initial state (the term above) includes two 'load' instances and four 'fault' instances that lead to markings with identical normal forms. Consequently, the combined rates are $2 \cdot 0.5$ and $4 \cdot 0.001$. Equivalent rewrites of the net structure are observed when $N > 2$.

5.1 Experimental Evidence

We conclude by showcasing the experimental validation of the method alongside a straightforward demonstration for calculating standard performance metrics. The results of the final-state location command are shown in Table 1 (above). This was carried out using Linux WSL on an 11th-gen Intel Core i5 with 40GB RAM. The state spaces align with those of the corresponding lumped CTMC. It is evident that analysis of large models is achievable by leveraging the model's symmetry. Note that the number of absorbing states in the TS quotient remains unchanged with N. Although a redundant state representation was used to construct the lumped CTMC directly, the efficiency of the `Maude` rewriting engine allowed us to estimate a time overhead of no more than 80%.

According to [9], the performance of modular RwPT was evaluated against symmetric nets (SN, previously referred to as well-formed nets) [11], which are colored Petri nets that produce a symbolic reachability graph (SRG) comparable (in its stochastic extension) to a lumped CTMC. As the values of N and K rise, the state aggregation level in modular RwSPT significantly surpasses that of SN. For instance, when $N = 10$, $K = 3$, and $M = 3$, the state aggregation level is about 45 times higher than SN. Furthermore, when $N = 10$, $K = 4$, and $M = 3$, it is roughly 200 times higher. This is due to the inherent hierarchical symmetry captured by modular RwPT, in contrast to the horizontal symmetry identified by SRG. You can replicate the experiments following the guidelines at https://github.com/lgcapra/rewpt/tree/main/modSPT/readme.

Table 1. Ordinary vs Quotient TS of `NPLsys(N,2,2)` † search timed out after 10 h

N	Ordinary		Quotient	
	states(final)	time (sec)	states(final)	time (sec)
1	60(2)	0	42(2)	0
2	779(4)	0.1	295(2)	0.1
3	6101(6)	4.8	1059(2)	0.9
4	37934(8)	69	2764(2)	3.6
5	204362(10)	818	5970(2)	10
6	1000187(12)	13930	11367(2)	47
7	-	†	19775(2)	85
10	-	†	73215(2)	2450

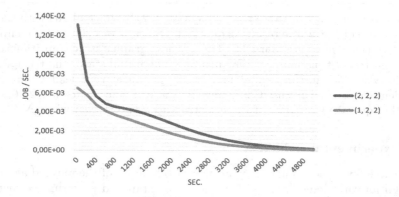

Fig. 3. System Throughput.

Figure 3 shows the system throughput, while Fig. 4 shows its reliability as a time function. As expected, both metrics decrease with time; additionally, the scenario that involves more replicas demonstrates increased throughput and enhanced reliability. To evaluate the system's performance, Fig. 5 shows the throughput while the system is operational, which is the ratio between the graphs in Figs. 3 and 4. It can be seen that the throughput is close to that of a single line, which, given the parameters, is $1/202.5 = 4.98E - 03$. The inflection point at around time 800 in both curves represents the system's reconfiguration time. The increased execution time of the job is a result of a system failure.

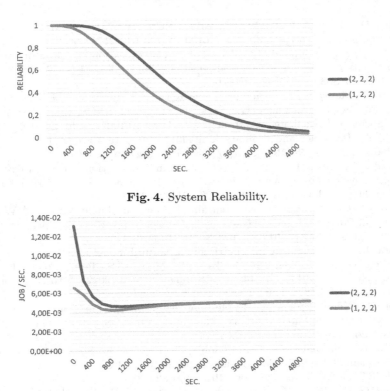

Fig. 4. System Reliability.

Fig. 5. System Throughput conditioned to its reliability.

6 Conclusion and Future Work

We have established a Lumped Markov process for modular, rewritable stochastic Petri nets (RwPT), a potent methodology for examining adaptive distributed systems encoded in Maude. RwPT models, constructed and manipulated with a concise set of (algebraic) operators, highlight structural symmetries leading to an efficient quotient state transition graph. Through an example of a gracefully degrading system, we have sketched a semi-automatic procedure for extracting the CTMC infinitesimal generator from the RwPT quotient graph.

References

1. Babai, L.: Graph isomorphism in quasipolynomial time [extended abstract]. In: Proceedings of the Forty-Eighth Annual ACM Symposium on Theory of Computing, STOC 2016, pp. 684–697. Association for Computing Machinery, New York (2016). https://doi.org/10.1145/2897518.2897542
2. Barbierato, E., Gribaudo, M., Iacono, M., Marrone, S.: Performability modeling of exceptions-aware systems in multiformalism tools. In: Al-Begain, K., Balsamo, S.,

Fiems, D., Marin, A. (eds.) ASMTA 2011. LNCS, vol. 6751, pp. 257–272. Springer, Heidelberg (2011). https://doi.org/10.1007/978-3-642-21713-5_19
3. Bouhoula, A., Jouannaud, J.P., Meseguer, J.: Specification and proof in membership equational logic. Theoret. Comput. Sci. **236**(1), 35–132 (2000). https://doi.org/10.1016/S0304-3975(99)00206-6
4. Bruni, R., Meseguer, J.: Generalized rewrite theories. In: Baeten, J.C.M., Lenstra, J.K., Parrow, J., Woeginger, G.J. (eds.) ICALP 2003. LNCS, vol. 2719, pp. 252–266. Springer, Heidelberg (2003). https://doi.org/10.1007/3-540-45061-0_22
5. Buchholz, P.: Exact and ordinary lumpability in finite Markov chains. J. Appl. Probab. **31**(1), 59–75 (1994). http://www.jstor.org/stable/3215235
6. Camilli, M., Capra, L.: Formal specification and verification of decentralized self-adaptive systems using symmetric nets. Discrete Event Dyn. Syst. **31**(4), 609–657 (2021). https://doi.org/10.1007/s10626-021-00343-3
7. Capra, L.: Canonization of reconfigurable PT nets in Maude. In: Lin, A.W., Zetzsche, G., Potapov, I. (eds.) Reachability Problems, pp. 160–177. Springer, Cham (2022). https://doi.org/10.1007/978-3-031-19135-0_11
8. Capra, L.: Rewriting logic and petri nets: a natural model for reconfigurable distributed systems. In: Bapi, R., Kulkarni, S., Mohalik, S., Peri, S. (eds.) ICDCIT 2022. LNCS, vol. 13145, pp. 140–156. Springer, Cham (2022). https://doi.org/10.1007/978-3-030-94876-4_9
9. Capra, L., Köhler-Bußmeier, M.: Modular rewritable petri nets: an efficient model for dynamic distributed systems. Theoret. Comput. Sci. **990**, 114397 (2024). https://doi.org/10.1016/j.tcs.2024.114397
10. Chiola, G., Dutheillet, C., Franceschinis, G., Haddad, S.: Stochastic well-formed colored nets and symmetric modeling applications. IEEE Trans. Comput. **42**(11), 1343–1360 (1993). https://doi.org/10.1109/12.247838
11. Chiola, G., Dutheillet, C., Franceschinis, G., Haddad, S.: A symbolic reachability graph for coloured petri nets. Theoret. Comput. Sci. **176**(1), 39–65 (1997). https://doi.org/10.1016/S0304-3975(96)00010-2
12. Chiola, G., Marsan, M.A., Balbo, G., Conte, G.: Generalized stochastic Petri nets: a definition at the net level and its implications. IEEE Trans. Software Eng. **19**, 89–107 (1993)
13. Clavel, M., et al.: All about maude - a high-performance logical framework: how to specify, program, and verify systems in rewriting logic. Lecture Notes in Computer Science. Springer (2007). https://doi.org/10.1007/978-3-540-71999-1
14. Iacono, M., Barbierato, E., Gribaudo, M.: The simthesys multiformalism modeling framework. Comput. Math. Appl. **64**(12), 3828–3839 (2012). https://doi.org/10.1016/J.CAMWA.2012.03.009
15. Iacono, M., Gribaudo, M.: Element based semantics in multi formalism performance models. In: MASCOTS 2010, 18th Annual IEEE/ACM International Symposium on Modeling, Analysis and Simulation of Computer and Telecommunication Systems, Miami, Florida, USA, 17–19 August 2010, pp. 413–416. IEEE Computer Society (2010). https://doi.org/10.1109/MASCOTS.2010.54
16. Padberg, J., Schulz, A.: Model checking reconfigurable petri nets with maude. In: Echahed, R., Minas, M. (eds.) ICGT 2016. LNCS, vol. 9761, pp. 54–70. Springer, Cham (2016). https://doi.org/10.1007/978-3-319-40530-8_4
17. Reisig, W.: Petri Nets: An Introduction. Springer, New York (1985). https://doi.org/10.1007/978-3-642-69968-9

Analytical Modelling of Asymmetric Multi-core Servers

M. Gribaudo[1](✉) and T. Phung-Duc[2]

[1] Politecnico di Milano, Milano, Italy
marco.gribaudo@polimi.it
[2] University of Tsukuba, Tsukuba, Japan

Abstract. Asymmetric Multi-core systems are the current way in which both embedded systems and data-centers are organized to obtain the best trade-off between the conflicting goals of saving energy and obtaining high performances. The idea is to have two different execution environments, capable of running exactly the same set of applications, but one tailored for energy efficiency at the expense of performance, and the other fast but more demanding in terms of resources. Such systems become even more interesting when combined with intermittent renewable energy sources, such as solar panels, that can reduce the impact of high performance cores, increasing the opportunities for their usage. In this work we will present a simple, yet effective, queuing network model that can expose and explain many of the critical aspects of these systems.

Keywords: Performance evaluation · Data Center · Asymmetric CPUs · Energy Efficiency

1 Introduction

Nowadays, ICT (Information and Communication Technology) is indispensable in our society and ICT is build in various systems from micro to macro ones. These ICT systems also consume a large amount of energy, and their workload is time dependent: it is high during the working hours, while it decreases during nighttime. The Asymmetric Multi-core processors (AMPs) or systems are proposed as a solution to cope with the problems mentioned above. From a system point of view, we also observe the same phenomenon in data centers where the traffic has a peak-on and peak-off nature. Motivated by these applications, we present in this paper several queuing systems which reflect key features of AMPs as well as power-saving data centers. On the other hand, energy consumption is not the only issue that ICT supported applications must face: many of the scenario where value is extracted from data, such as Big Data applications [1], require answers to be produced in relatively short amount of time [2]. The tools presented in this work, can be used to find the best tradeoff between energy consumption and performance.

This paper is organized as follows. In Sect. 2, we present in detail some related work. Section 3, the asymmetric multi-core servers are presented and in Sect. 4, queueing models with and without energy-harvesting for these systems are presented. Performance evaluation results are presented in Sect. 5.

2 Related Work

Nowadays, data centers are the core infrastructure of our information era. On the other hand, data centers are also responsible for a large portion of carbon emissions, since they consume a huge amount of energy. However, not all the servers in data centers are fully operated due to the peak-on and peak-off nature of Internet's traffic [3]: even when the server (or a core) is idle not serving a job, it still consumes about 60–70% of energy while processing a job [3]. Thus, studies on turning on/off servers to save energy have been extensively carried out, recently. One of the problems in these On/Off policies is that an off server cannot be active immediately when jobs arrive. In order to be active serving a job, a server needs some setup time during which it cannot serve a job but consumes quite a large amount of energy [4]. Thus, in some situation where the traffic intensity is relatively large, it is better to keep all the servers on forever [5]. Mitrani [6,7] proposed some models where a group of servers is turned on and off according to thresholds on the number of jobs in the system. Algorithms for calculating the thresholds were presented in [8]. Related models with energy harvesting or energy packets can be found in [9,10].

Motivated by this situation, a natural idea is that we use a good mix of both high and low performance servers. These servers are kept on all the time so they can immediately serve a job. It is clear that a high performance server (core) can process at high speed but also consumes more energy. On the other hand, low performance server can process only at slow speed but also consumes less energy. However, as long as these servers are idle, they can serve jobs immediately and thus do not need setup times. This is similar to the multi-core architecture presented in Sect. 3. Our research question is the following "does the asymmetric core architecture outperform the symmetry one?".

In order to answer this research question, we propose four queueing models with heterogeneous servers for both asymmetric multi-core CPUs, as well as for power-saving data centers. In our models, there are two types of servers with low and high service rates. We consider several cases where a jobs in slow servers can or cannot be migrated immediately to fast servers once they are available. The former case is easily analyzed using a birth-death process while the latter needs a quasi-birth-and-death formulation.

Performance evaluation for asymmetric multi-core systems has been extensively studied [11–14]. In these studies, power-consumption was the main performance measure and waiting time or queue-length was not studied. However, in reality both power-consumption and queuing performance should be considered concurrently because these two performance measures have a trade-off relation. A scenario where servers using different technologies, for example Intel or ARM, was also considered in [15]. Models with heterogeneous servers of different service rates are also considered in [16,17] for 5G systems.

3 Asymmetrical Multi-core Servers

The type of systems we consider in this work are typical implemented in two main different flavours: Big-Little multi-core CPU architectures and data centers equipped with both performance oriented and energy efficient of servers.

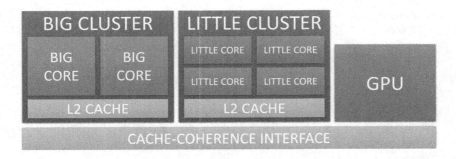

Fig. 1. The architecture of a typical asymmetrical Big-Little multi core CPU.

The architecture of a typical Big-Little architecture is shown in Fig. 1. The CPU is composed of two clusters of cores: the "Big" ones usually occupy a larger area on the processor, and can offer a better performance due to the way in which their pipeline is implemented and unrolled; the "Little" cores are grouped in another cluster, and due to their smaller footprint, are generally more numerous. Each cluster has its own L2 cache, which must be synchronized with the one of the other using an important cache-coherence interface. This component is usually the bottleneck in applications that migrate from one cluster to the other, limiting the applicability of this feature. In some software / hardware architectures, migration is not even possible, and the type of cores must be selected at the beginning of the computation. Usually, this type of CPU is also equipped with a GPU, which however is not considered in this work.

The infrastructure, of the typical data-center considered in this work, is shown in Fig. 2. In this case, the servers, which are usually distributed into corridors, with hot and cold aisles to allow a better cooling of the devices, can be grouped into two clusters: *High Performance Servers*, and *Energy Efficient Servers*. The two types of nodes can differ in terms of CPU, GPU or other accelerator units, disks and network devices. In some cases, CPUs can even have different architectures, like ARM or Intel, and migration might only be possible when specific types of software / storage infrastructures are employed.

4 Queues with Asymmetrical Servers

In this section, we present several queueing models for AMPs. To keep the discussion as general as possible, we will not refer to the processors as fast or slow, energy efficient or power intensive, but simply as "type 1" and "type 2". Figure 3 represents a model with n_1 servers of type 1 with service rate μ_1 and n_2 servers of type 2 with service rate μ_2, modulated by two states which alternate at rates γ_1 and γ_2, which are depicted as a sun and a cloud icon.

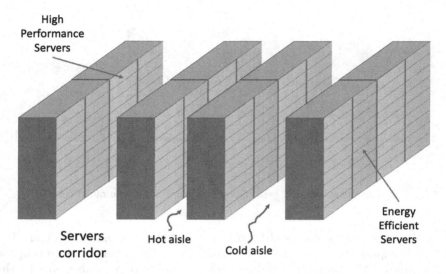

Fig. 2. The architecture of a data-center employing both performance-oriented and energy-efficiency-oriented servers.

We consider several cases where jobs can or cannot migrate from one type of server to the other one. Jobs arrive a the system according to a Poisson process with rate λ. In this model, servers of type 1 are used first, meaning that unless all servers of type 1 are fully occupied, a job will be allocated to an idle server of type 1. If all servers of type 1 are occupied, the job will be allocated to a type 2 server. In case all servers are busy, the job is placed in the buffer. When a job completes, two different scenarios are possible. With the *Migration* approach, when a job in a type 1 server completes, a job from a type 2 server (if present) migrates to occupy its position, and the type 2 server is occupied by a waiting job in the buffer if any. Due to migration, the number of jobs in the system forms a Birth-and-Dearth process as shown in Fig. 4.

The second scenario, does not allow migration, as shown in Fig. 5. In this model, a job is served by the same server until departure. In this case, we need a two-dimensional Markov chain representing the number of busy servers of type 1 and of type 2, whenever the number of jobs in the system is not greater than

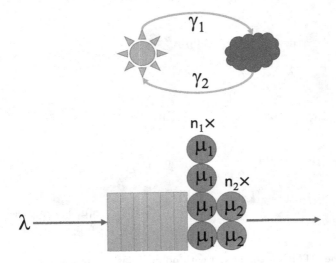

Fig. 3. The proposed Markov-modulated multiple-server queuing station, with asymmetrical cores.

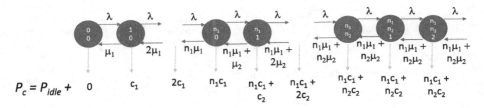

Fig. 4. The CTMC of the model with no modulation, and server migration.

(n_1+n_2). When the number of jobs in the system is greater than (n_1+n_2), only a single dimension (birth-death-process) is needed as n_1 jobs are served by the type 1 servers and n_2 jobs are served by the type 2 servers, while the remaining jobs are waiting at the buffer. The departure rate in these states (death rate) is constant $(n_1\mu_1 + n_2\mu_2)$, as shown in Fig. 5.

Figure 6 shows the transition diagram for the case of modulation and migration. In this case, the energy is harvested from the environment. When the environment is good (representing by green circles), the servers of type 1 are prioritized to be used first. Once the environment changes to a bad state (represented in orange) servers of type 2 are given priority. We assume that jobs can be migrated from one server type to the other. We can observe that Fig. 6 consists of two birth-and-death processes which are interchanged according to the state of the environment. It should be noted that the exact transition rates in Fig. 6 depend on the values of n_1 and n_2, since migration might not be possible for some jobs when $n_1 \neq n_2$ due to the unavailability of servers of the different type.

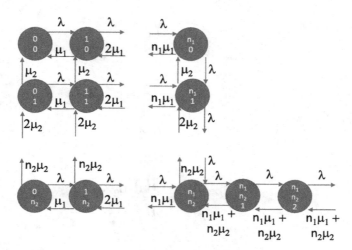

Fig. 5. The CTMC of the model with no modulation, and no server migration.

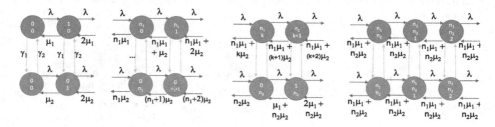

Fig. 6. The CTMC of the model with modulation, and server migration.

Finally, Fig. 7 shows the most complex case, where there is modulation without migration. We can see that Fig. 7 is a combination of two models described in Figs. 5, with interchanged rates, which are swapped according to the state of the environment. The main difference of the two blocks occurs when the number of jobs in the system is less than $\min(n_1, n_2)$. In this case, depending on the macro block, an arrival will cause either a horizontal or a vertical transition: a red arrow emphasizes the main transition direction in which arrivals are routed starting from an empty system.

4.1 Evaluation

In this section we describe the performance metrics of the presented systems are evaluated. First, for the transition diagram in Fig. 4, the stationary distribution is easily obtained because it is simply a birth-death-process with homogeneous structure for states with at least $n_1 + n_2$ jobs. From the stationary distribution, we compute the mean sojourn time (response time) of jobs and the mean power consumption P_c.

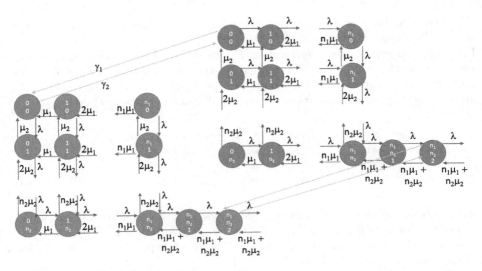

Fig. 7. The CTMC of the model with modulation, and no server migration.

In particular, let as call $P(i, j, k)$ the probability of having i jobs of type 1, j jobs of type 2 in service, and k jobs in the queue. Due to the particular structure of this model, $j > 0 \implies i = n_1$, and $k > 0 \implies i = n_1 \wedge j = n_2$. We then have

$$P(i, 0, 0) = p_0 \cdot \frac{\lambda^i}{i! \, \mu_1^i}$$

$$P(n_1, j, 0) = p_0 \cdot \frac{\lambda^{n_1}}{n_1! \, \mu_1^{n_1}} \cdot \prod_{l=1}^{j} \left(\frac{\lambda}{n_1 \cdot \mu_1 + l \cdot \mu_2} \right)$$

$$P(n_1, n_2, k) = p_0 \cdot \frac{\lambda^{n_1}}{n_1! \, \mu_1^{n_1}} \cdot \left(\prod_{l=1}^{n_2} \left(\frac{\lambda}{n_1 \cdot \mu_1 + l \cdot \mu_2} \right) \right) \cdot \left(\frac{\lambda}{n_1 \cdot \mu_1 + n_2 \cdot \mu_2} \right)^k \quad (1)$$

with $p_0 = p(0, 0, 0)$ chosen such that:

$$p_0 + \sum_{i=1}^{n_1} p(i, 0, 0) + \sum_{j=1}^{n_2} p(n_1, j, 0) + \sum_{k=1}^{\infty} p(n_1, n_2, k) = 1 \quad (2)$$

Although the summation in the last term goes to infinity, this can be computed since:

$$\sum_{k=1}^{\infty} p(n_1, n_2, k) = \frac{p(n_1, n_2, 1)}{1 - \frac{\lambda}{n_1 \cdot \mu_1 + n_2 \cdot \mu_2}} \quad (3)$$

The average number of jobs in the system N can be computed as:

$$N = \sum_{i=1}^{n_1} i \cdot p(i, 0, 0) + \sum_{j=1}^{n_2} (n_1 + j) \cdot p(n_1, j, 0) + \sum_{k=1}^{\infty} (n_1 + n_2 + k) \cdot p(n_1, n_2, k) \quad (4)$$

We then used Eq. 4 together with Little's law, and we determined the average System Response Time $R = N/\lambda$.

For what concerns power consumption P_c, we ignore the idle consumption and we focus on the component determined by the utilization of the resource. In particular, we suppose that each node of type 1 and type 2 cause respectively an increase of power consumption c_1 and c_2. We then compute P_c as:

$$P_c = \sum_{i=1}^{n_1} i \cdot c_1 \cdot p(i,0,0) + \sum_{j=1}^{n_2}(n_1 \cdot c_1 + j \cdot c_2) \cdot p(n_1,j,0) + \sum_{k=1}^{\infty}(n_1 \cdot c_1 + n_2 \cdot c_2) \cdot p(n_1,n_2,k) \quad (5)$$

Although an explicit closed form expression, with a finite number of terms, could be derived for both Eq. 4 and Eq. 5, due to the speed at which $p(n_1, n_2, k) \to 0$ as $k \to \infty$ in the considered scenario, we compute both expressions via truncation of the solution at a suitably large value of k.

The model from Fig. 5 has a birth-and-death structure for the states with at least $n_1 + n_2$ jobs and a two-dimensional one for states with less than or equal to $n_1 + n_2$ jobs. In this case we only have the following restriction on states, $k > 0 \implies i = n_1 \wedge j = n_2$, since queueing can only take place when both Types of servers are full. Since the birth-and-death part can be solved explicitly, we resort to finding the steady-state probabilities for states in the part with less than $n_1 + n_2$ jobs in the system. In particular, following [18] we create a $(n_1 + 1) \times (n_2 + 1)$ defective infinitesimal generator matrix $\mathbf{B_{00}}$, whose elements corresponds to the probabilities $p(i,j,0)$ of having $0 \leq i \leq n_1$ jobs served by nodes of type one, and $0 \leq j \leq n_2$ of type one. Let us call $q_{[i,j][i',j']}$ the elements of $\mathbf{B_{00}}$ corresponding to a jump from a state with i jobs served by type 1 and j by type 2, to a state with i' and j' jobs of the two types. We have:

$$q_{[i,j][i',j']} = \begin{cases} \lambda & \text{if } i' = i-1 \wedge i < n_1 \\ \lambda & \text{if } j' = j-1 \wedge i = n_1 \wedge i < n_2 \\ i \cdot \mu_1 & \text{if } i' = i+1 \wedge i > 0 \\ j \cdot \mu_2 & \text{if } j' = j+1 \wedge j > 0 \\ 0 & \text{otherwise} \end{cases} \quad (6)$$

The element in the diagonal, $q[i,j][i,j]$ is defined as usual as negative sum of the elements in the row:

$$q[i,j][i,j] = - \sum_{[i',j'] \neq [i,j]} q[i,j][i',j'] \quad (7)$$

except for the element corresponding to $[n_1, n_2]$ that is:

$$q[n_1,n_2][n_1,n_2] = -\lambda - \sum_{[i',j'] \neq [n_1,n_2]} q[i,j][i,j] \quad (8)$$

which makes matrix $\mathbf{B_{00}}$ defective. We then define the column vector $\mathbf{b_{01}}$ with all the elements equal to zero, except the one corresponding to state $[i,j]$, which is set to λ, and row vector $\mathbf{b}_{10}$ again with all the elements equal to zero, except for the one corresponding to state $[i,j]$ that is set to $n_1 \cdot \mu_1 + n_2 \cdot \mu_2$. We also set the last diagonal element $b_{11} = -n_1 \cdot \mu_1 - n_2 \cdot \mu_2$. These matrices are arranged in a $(n_1 + 1) \cdot (n_2 + 1) + 1$ square matrix $\mathbf{Q}$:

$$\mathbf{Q} = \begin{array}{|c|c|} \hline \mathbf{B}_{00} & \mathbf{b}_{01} \\ \hline \mathbf{b}_{10} & b_{11} \\ \hline \end{array} \qquad (9)$$

This is then used to find a solution of the following matrix equation with the given normalizing condition:

$$\begin{cases} |\ldots p(i,j,0) \ldots, p(n_1, n_2, 1)| \cdot \mathbf{Q} = \mathbf{0} \\ \sum_{i=0}^{n_1} \sum_{j=0}^{n_2} p(i,j,0) + \dfrac{p(n_1, n_2, 1)}{1-\rho} = 1 \end{cases} \qquad (10)$$

with:

$$\rho = \frac{\lambda}{n_1 \cdot \mu_1 + n_2 \cdot \mu_2} \qquad (11)$$

In this context, the probability of having $k > n_1 + n_2$ jobs in queue becomes:

$$p(n_1, n_2, k) = p(n_1, n_2, 1)\rho^{k-1} \qquad (12)$$

Performance metrics can be computed again using a slight modification of Eq. 4 and Eq. 5:

$$N = \sum_{i=0}^{n_1} \sum_{j=0}^{n_2} (i+j) \cdot p(i,j,0) + \sum_{k=1}^{\infty} (n_1 + n_2 + k) \cdot p(n_1, n_2, k) \qquad (13)$$

$$P_c = \sum_{i=0}^{n_1} \sum_{j=0}^{n_2} (i \cdot c_1 + j \cdot c_2) \cdot p(i,j,0) +$$

$$(n_1 \cdot c_1 + n_2 \cdot c_2) \cdot \sum_{k=1}^{\infty} p(n_1, n_2, k) \qquad (14)$$

The transition diagram in Fig. 6 exhibits a Quasi-birth-and-death structure, for which the stationary distribution can be obtained in a similar way. In this case, however, $\mathbf{B}_{01}$ and $\mathbf{B}_{10}$ will be respectively a $2(n_1+n_2)$ and a $2(n_1+n_2) \times 2$ matrix. In particular, they are defined as follows. First we have to note that the system state, when all the servers are not full, can be in three different configurations. To simplify the discussion, let us suppose that $n_2 = n_1 + E$, with $E > 1$ (i.e. we have strictly more type 2 servers than type 1). If in the system there are a total of n jobs, and $n \leq n_1$, whenever the modulating process switches, jobs are simply transferred from one type to the other. If we have $n_1 < n < n_2$ jobs in the system, when the modulating process switches, we jump from a state with $i = n_1, j = n - n_2$ to one with $i = 0, j = n$, where i and j represent respectively the number of servers of type 1 and type 2. Finally, if $n_2 < j \leq n_1 + n_2$, we jump from a state with $i = n_1, j = n - n_2$ to one with

$i = n - n_1, j = n_2$. For space constraints, we skip the precise definition of matrix **Q**, which can be derived directly from Fig. 6.

In addition, we do not have a simple term ρ to characterize the states with queue $k > 0$, but a matrix **R**, solution of the following equation:

$$\mathbf{A}_0 + \mathbf{R} \cdot \mathbf{A}_1 + \mathbf{R}^2 \cdot \mathbf{A}_2 = 0 \qquad (15)$$

with:

$$\mathbf{A}_0 = \begin{vmatrix} \lambda & 0 \\ 0 & \lambda \end{vmatrix} \quad \mathbf{A}_1 = \begin{vmatrix} -\lambda - \mu_T - \gamma_1 & \gamma_1 \\ \gamma_2 & -\lambda - \mu_T - \gamma_2 \end{vmatrix} \quad \mathbf{A}_2 = \begin{vmatrix} \mu_T & 0 \\ 0 & \mu_T \end{vmatrix} \qquad (16)$$

where

$$\mu_T = n_1 \cdot \mu_1 + n_2 \cdot \mu_2 \qquad (17)$$

and the probability of having k jobs defined as:

$$p(n_1, n_2, k) = \mathbf{p}(n_1, n_2, 1) \cdot \mathbf{R}^{(k-1)} \cdot \begin{vmatrix} 1 \\ 1 \end{vmatrix} \qquad (18)$$

The rate matrix R can be calculated using various existing algorithms [18].

Because in states with more than $(n_1 + n_2)$ jobs, the transition structure is homogeneous, the generating function approach is also convenient to find the stationary probabilities as well as the mean response time. However, because all elements of matrix $\mathbf{A}_1$ are non-zero, we cannot have explicit solutions for the stationary distribution as in [19] via the generating function method. From the stationary probabilities, we can evaluate the power consumption of the system, in a way similar to what was presented earlier in this section.

Finally, the transition diagram in Fig. 7 can also be solved with a Matrix Analytical technique: in this case, it exhibits the same regular structure as the previous case, but a more complex initial part. So basically, we can obtain the stationary distributions for all four models and calculate the performance measures such as mean response time and mean power-consumption. These performance measures can be used to optimize the design of the system, allowing to choose the best number of type 1 and type 2 servers, for a particular scenario.

5 Results

Let us first focus on the case in which we do not take into account renewable sources. Figure 8 shows the queue length distribution when there are $n_1 = 8$ fast machines and $n_2 = 16$ slow machines, where the fast ones are two time faster than the slow ones and the total system capacity is $n_1\mu_1 + n_2\mu_2 = 16$ job per second, for three different arrival rates $\lambda \in [1.5, 8.5, 15.5]$ jobs per sec. In Fig. 8a) fast nodes are chosen first, while in Fig. 8b) priority is given to the slow ones. As expected, giving priority to the fast nodes reduces the average queue length for low loads, while in high load regimes the differences between the two priority schemes become negligible.

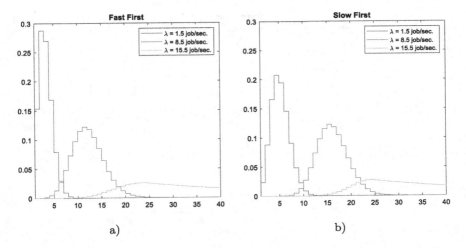

Fig. 8. Distribution of the number of jobs in the queue: a) priority to performance node, b) priority to efficiency nodes.

We then study the impact on the number of available resources. We keep the total computational power of the system constant, and we change the proportion α between machines of type 1, with respect to the the total number of nodes, that is: $\alpha = \frac{n_1}{n_1+n_2}$. Figure 9 shows the average queue length, while considering $n_1 + n_2 = 24$, and $n_1\mu_1 + n_2\mu_2 = 16$ jobs per second, varying the arrival rate λ, ranging from the case $n_1 = 4$ and $n_2 = 20$, to the case $n_1 = 20$ and $n_2 = 4$. Priority is always given to nodes of type 1, which works two times faster than the servers of type 2. Please note that since the total capacity of the system is fixed, each scenario will have different values for the service rates of both types of nodes, μ_1 and μ_2. As expected, having a large number of fast nodes reduces the queue when this system load is low. However, to maintain a constant total capacity of the system, their actual speed is slower with respect to the case in which there is a larger number of type 2 nodes. This becomes evident as the load increases: when the load is very high, it becomes better to have a larger number of type 2 nodes to better handle the extra demand.

We then study the average queue length when we keep the number of resources of type 1 fixed , (in particular $n_1 = 8$), and we increase the resources of type 2. Again, the total capacity is fixed to $n_1\mu_1 + n_2\mu_2 = 16$ jobs per second, and type 1 servers are considered to be two times faster than the ones of type 2, which means the actual speed of each individual server decreases with the growth of n_2. Figure 10 shows that the effect of increasing the total number of cores, keeping the total capacity constant, has a negative impact on performance, because it reduces the speed of the available resources, letting the increased number of servers have an impact only when the total population of the system raises.

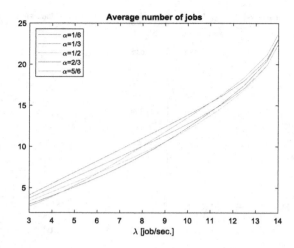

Fig. 9. Average number of jobs in the queue, for different proportions of the considered servers α.

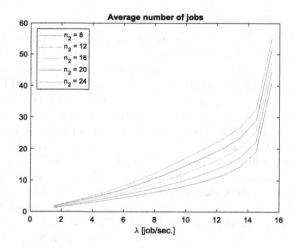

Fig. 10. Effect of increasing n_2, the number of servers of type 2, to the number of jobs in the queue.

We finally fix the number of resources to $n_1 = 8$ plus $n_2 = 16$, and we study the impact on the speed difference. In particular, let us call β the speedup of servers of type 1 with respect to the one of type 2, that is: $\mu_1 = \beta\mu_2$.

Again, the total service rate of the system is kept constant. Figure 11a) shows the average queue length, for $\beta \in [1.25, 3]$. The impact of faster servers is not too evident, even if the use of faster type 1 machines tends to provide better performance. However this comes at a cost: as Fig. 11b) shows, a small increase of performances is matched by a much larger increase in energy consumption.

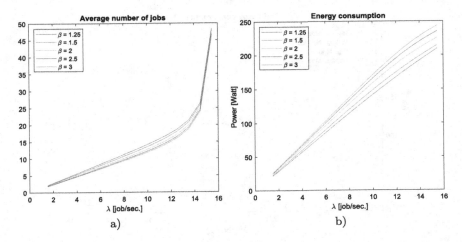

Fig. 11. Effect of increasing β, the speed of servers of type 1, with respect to the one of type 2: a) average number of jobs, b) energy consumption.

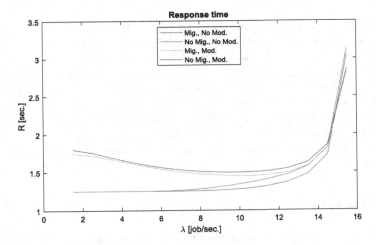

Fig. 12. System Response time for the considered scenario under variable load.

Energy consumption is computed ignoring the idle term, with a maximum power consumption that is proportional to the square of the server' speed, that is: $c_i = 20\,\mu_i^2$ W.

5.1 Renewable Sources

We now take into account the use of renewable sources, which changes the priority in selecting resources from type 1 to type 2. Figure 12 shows the evolution of the response time, as function of the arrival rate, for a case with $n_1 = 8$

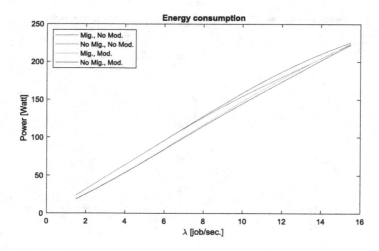

Fig. 13. Power Consumption for the considered scenario under variable load.

and $n_2 = 16$, with $\mu_1 = 0.8$ jobs/sec. and $\mu_2 = 0.4$ job/sec. for the four different cases. The renewable energy source alternates at speed $\gamma_1 = 0.1$ sec.$^{-1}$ and $\gamma_2 = 0.125$ sec.$^{-1}$, and it is supposed to be powerful enough to completely sustain the functioning of the servers. The energy consumption in Watt (W) is shown in Fig. 13, where the maximum power absorbed by one core is respectively $c_1 = 12.8$ W and $c_2 = 3.2$ W. As seen in Figs. 12 and 13, taking into account the availability of the renewable energy source produces a non-monotonic behaviour in the response time, but allows a reduction in the power consumption. Instead, migration reduces the response time, at the expense of an increased energy consumption.

6 Conclusion

Although the considered model is very simple, it has some interesting features, that makes the considered system quite unique. Indeed asymmetric systems can provide a good trade-off between performance and energy consumption, and the ability to take into account the availability of renewable sources can improve the effectiveness of the technique: simple modelling results such as the ones proposed in this paper can help to fine-tune the system and choose the more appropriate configurations to achieve a given objective.

Future work will consider more classes of jobs and decide which type of servers is assigned based on the this extra information. Moreover, validation using simple edge computing based implementations will provide a better assessment of the considered scenario.

References

1. Gandini, A., Gribaudo, M., Knottenbelt, W.J., Osman, R., Piazzolla, P.: Performance evaluation of NoSQL databases. In: Horváth, A., Wolter, K. (eds.) EPEW 2014. LNCS, vol. 8721, pp. 16–29. Springer, Cham (2014). https://doi.org/10.1007/978-3-319-10885-8_2
2. Barbierato, E., Gribaudo, M., Iacono, M.: A performance modeling language for big data architectures. In: ECMS 2013, pp. 511–517 (2013)
3. Barroso, L.A., Hölzle, U.: The case for energy-proportional computing. Computer **40**(12), 33–37 (2007)
4. Gandhi, A., Doroudi, S., Harchol-Balter, M., Scheller-Wolf, A.: Exact analysis of the M/M/k/setup class of Markov chains via recursive renewal reward, in: Proceedings of the ACM SIGMETRICS/International Conference on Measurement and Modeling of Computer Systems, pp. 153–166 (2013)
5. Phung-Duc, T.: Exact solutions for M/M/c/setup queues. Telecommun. Syst. **64**, 309–324 (2017)
6. Mitrani, I.: Service center trade-offs between customer impatience and power consumption. Perform. Eval. **68**, 1222–1231 (2011)
7. Mitrani, I.: Managing performance and power consumption in a server farm. Ann. Oper. Res. **202**, 121–134 (2013)
8. Tournaire, T., Castel-Taleb, H., Hyon, E.: Service center trade-offs between customer impatience and power consumption. ACM Trans. Model. Perform. Eval. Comput. Syst. **8**, 1–31 (2023)
9. Politaki, D., Alouf, S.: Stochastic models for solar power. In: Reinecke, P., Di Marco, A. (eds.) EPEW 2017. LNCS, vol. 10497, pp. 282–297. Springer, Cham (2017). https://doi.org/10.1007/978-3-319-66583-2_18
10. Gelenbe, E.: A sensor node with energy harvesting. ACM SIGMETRICS Perform. Eval. Rev. **42**, 37–39 (2014)
11. Gomatheeshwari, B., Selvakumar, J.: Appropriate allocation of workloads on performance asymmetric multicore architectures via deep learning algorithms. Microprocess. Microsyst. **73**, 102996 (2020)
12. Balakrishnan, S., Rajwar, R., Upton, M., Lai, K.: The impact of performance asymmetry in emerging multicore architectures. In: In 32nd International Symposium on Computer Architecture (ISCA 2005), pp. 506–517. IEEE. (2005)
13. Mittal, S.: A survey of techniques for architecting and managing asymmetric multicore processors. ACM Comput. Surv. (CSUR) **48**, 1–38 (2016)
14. Pricopi, M., Muthukaruppan, T.S., Venkataramani, V., Mitra, T., Vishin, S.: Power-performance modeling on asymmetric multi-cores. In: 2013 International Conference on Compilers, Architecture and Synthesis for Embedded Systems (CASES), pp. 1-10. IEEE. (2013)
15. Barbierato, E., Manini, D., Gribaudo, M.: A multiformalism-based model for performance evaluation of green data centres Electronics **12**(10), 2169 (2023)
16. Ren, Y., Phung-Duc, T., Liu, Y.K., Chen, J.C., Lin, Y.H.: ASA: Adaptive VNF scaling algorithm for 5G mobile network. In: 2018 IEEE 7th International Conference on Cloud Networking (CloudNet), pp. 1–4 (2018)
17. Sato, M., Kawamura, K., Kawanishi, K., Phung-Duc, T.: Modeling and performance analysis of hybrid systems by queues with setup time, Performance Evaluation 162 (2023)

18. Neuts, M.F.: Matrix-analytic methods in queuing theory. Eur. J. Oper. Res. **15**(1), 2–12 (1984)
19. Servi, L.D., Finn, S.G.: M/M/1 queues with working vacations (M/M/1/WV). Perform. Eval. **50**(1), 41–52 (2002)

Robust Streaming Benchmark Design in the Presence of Backpressure

Iain Dixon(✉), Matthew Forshaw, and Joe Matthews

Newcastle University, Newcastle upon Tyne NE4 5TG, UK
{iain.dixon,matthew.forshaw,joe.matthews}@ncl.ac.uk

Abstract. Replicability and reproducibility are critical challenges in systems research, particularly in evaluating systems that experience performance variability due to hardware and complex self-regulating behaviours. This paper investigates performance evaluation practices for stream processing systems, focusing on the impact of backpressure. Backpressure occurs when data is received faster than it can be processed, leading to cascading delays and potential data loss. Through empirical analysis, we demonstrate where popular closed-loop benchmark designs used in benchmarks such as NEXMark and YCSB under backpressure conditions may fail to meet target arrival rates, leading to unreliable benchmarking results. Our study provides recommendations for metrics to better understand system behavior, and proposes best practices for reliable performance evaluation in the presence of backpressure.

Keywords: Stream Processing · Benchmarking · Backpressure

1 Introduction

Replicability and reproducibility [34] are contemporary challenges in the area of systems research [37,40]. These challenges are greatest in the evaluation of systems which exhibit complex performance characteristics [8] and experience performance variability due to underlying hardware [12,22]. The potential deployment design space for these systems may also limit our ability to explore possible configurations within reasonable experimental cost [4]. There have been efforts in the literature to establish robust experimental methodologies [13,26], including ordering of experiments in cloud environments [7] and randomised multiple interleaved trials [1]. Nevertheless, ongoing scrutiny of benchmarking practices is essential.

In this study, we explore performance evaluation practices for stream processing systems (SPSs). A preliminary study of open-source benchmarks, including NEXMark [35] and YCSB [6], found common practices using closed-loop [30] workload generators to provide a more controllable benchmark environment. Inspired by previous works [27,36], e.g. Gil Tene's work on *coordinated omission*, which found interaction effects between the system under test and the benchmark, we explore the implications of closed-loop benchmark designs for SPSs.

We specifically evaluate the impact of a common flow control mechanism, *backpressure*, on benchmarking practices. Backpressure occurs when a stream processing system receives data faster than it can be processed, causing operators to slow or halt, and processing delays propagate through the pipeline to upstream operators. Through a systematic literature survey, we identify that while backpressure is a well-documented self-regulating feature of Flink, it is currently underrepresented in the performance evaluation practices of academic literature.

We empirically evaluate the impacts of backpressure and demonstrate that self-regulating mechanisms such as backpressure may be difficult to observe through conventional metrics such as end-to-end latency and throughput alone. We present the implications of the backpressure mechanism on the design of streaming benchmarks and suggest improvements to overcome present deficiencies which challenge the robustness and reliability of benchmarking efforts.

The remainder of the paper is organised as follows. We first introduce preliminaries of stream benchmarking and backpressure (§ 2), and present our systematic literature review of backpressure in performance evaluation of SPSs (§ 3). We present our model of benchmarking (§ 4) and empirically evaluate the impact of backpressure (§ 5). We introduce a mechanism to overcome these challenges (§ 6). We conclude by highlighting threats to validity for the work (§ 7) before concluding and motivating future work (§ 8).

2 Preliminaries

We first establish the required knowledge for the remainder of this work. We introduce a high-level model of streaming systems, introduce the benchmarking practices and considerations for streaming systems and introduce backpressure.

2.1 Streaming

Stream processing systems (SPS) comprise a pipeline of chained operators that ingest tuples from an upstream system, and process them sequentially until they reach a downstream sink. Stream tuples are comprised of the data to be processed and a timestamp representing the order in which data was collected and transmitted. The source and sink may be remote systems streaming and receiving data (e.g., Kafka, IoT sensors) or databases feeding and storing data. SPS performance is typically measured with end-to-end tuple latency and throughput to assess the speed and volume of data processed.

2.2 Stream Benchmarking

Benchmarking SPS requires a framework which can produce and consume tuples similar to a source and sink. SPS benchmarks are either open or closed-loop, but the prevailing implementations of popular benchmarks like NEXMark and YCSB are closed [15,18]. Closed-loop benchmarks do not require formal isolation of the generator and system under test [18,31], which makes implementations simpler to distribute and utilise. From our preliminary literature search, the dominant approach appears to be closed-loop; therefore, in this work, we look specifically at closed-loop stream benchmarks.

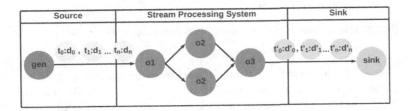

Fig. 1. SPS benchmark simulates an upstream source and downstream sink

Stream benchmarks are comprised of three parts: a generator that simulates the upstream source, the SPS under test, and a sink that simulates a downstream sink (as seen in Fig. 1). The generator creates streams of tuples and loads the SPS following an arrival process. The sink collects metrics on the output stream of tuples, which are then stored or discarded. This allows us to test the SPS while minimising the impact of network and contention characteristics a deployed SPS might experience.

2.3 Backpressure

Backpressure is a mechanism which throttles upstream operators to match the speed of an overburdened downstream operator (see Fig. 2). This effect propagates backwards through the pipeline until it reaches the ingestion operator. In deployment, backpressure results in load shedding, where incoming tuples that cannot be processed are dropped randomly [2].

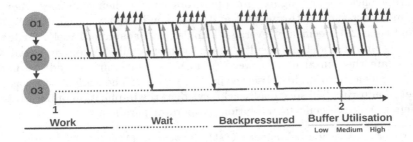

Fig. 2. Backpressure propagates through a deployed SPS and sheds load

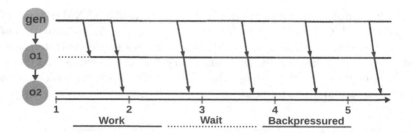

Fig. 3. Backpressure in closed-loop SPS can propagate to the generator

In a closed-loop stream benchmark, backpressure can propagate up to the generator and throttle the generation of new tuples (see Fig. 3). The mismatch of how SPS react to backpressure in deployment and benchmarking means that benchmark results may have no bearing on how an SPS performs in deployment.

3 Related Work

Backpressure flow control was first proposed in the context of packet network traffic by Tassiulas and Ephremides [33]. It has since been implemented in reactive systems such as Akka Streams and RxJava, as well as stream processing systems such as Storm, Spark Streaming, and Flink. Flink's backpressure mechanism is well documented within documentation and grey literature, yet it receives extremely limited attention in the literature.

Table 1 summarises our systematic literature review of backpressure within articles mentioning our target system, Apache Flink. We first conducted a search on Scopus for[1] all papers containing "Apache Flink" in the title, abstract, description or full text of papers, which returned 2,240 documents. A revised search of articles mentioning backpressure[2] returned only 36 (1.61%) documents.

Several papers mention the importance of backpressure outside of the experimental findings [5,10,11,16,19,21,28,39], but only a small minority account for backpressure within their experimental findings [20,23]. Several articles specifically address the performance implications of backpressure, but rather than presenting backpressure figures for operators within their pipelines, they use proxy measures such as throughput and end-to-end latency [3,9,15,17,29].

ContTune [20] specifically leverages Flink's `backPressuredTimeMsPerSecond` metric within its model and results. However, more could be done to specifically demonstrate the causal link between performance phenomena and the performance results and variability observed in their findings. The work of Ntoulias *et al.* on fleet monitoring [23] specifically acknowledges the implication of backpressure and demonstrates its effects in relation to operator parallelism.

---

[1] Query for all Apache Flink papers: (ALL ("apache flink")).
[2] Query for Flink papers mentioning Backpressure: (ALL ("apache flink") AND ALL ("backpressure" OR "back pressure" OR "back-pressure")).

Table 1. Analysis of Flink and backpressure literature by year and article type.

	Query One	Query Two				
Year	Total	Conference	Article	Book Chapter	Review	Total
2013	1	–	–	–	–	–
2015	21	–	–	–	–	–
2016	75	1	–	–	–	1 (1.33%)
2017	166	3	–	–	–	3 (1.81%)
2018	275	4	–	–	–	4 (1.45%)
2019	322	2	2	1	1	6 (1.86%)
2020	333	2	1	–	–	3 (0.9%)
2021	335	2	2	–	–	4 (1.19%)
2022	311	3	5	–	–	8 (2.57%)
2023	279	3	1	–	–	4 (1.43%)
2024	122	1	2	–	–	3 (2.46%)
All	2240	21	13	1	1	36 (1.61%)

4 Stream Generators

Stream benchmark generators have slight differences based on their implementation or the SPS use case. After analysing implementations of NEXMark and YCSB [6,15] we have abstracted the commonalities between them into the **Basic Loop** as seen in Algorithm 1. Stream benchmark generators create and transmit tuples following an arrival rate $\lambda = \frac{n}{\delta}$, where n tuples arrive every δ seconds.

Algorithm 1: Basic Loop
input: W, δ_L, λ

for $w := 0$ *to* $W - 1$ **do**
  $n_w := \lambda_w * \delta_L / 1000;$
  $t_{start} :=$ TIME();
  for $i := 0$ *to* $n_w - 1$ **do**
    tuple:=GEN();
    SEND(tuple);
  $\delta_{transmit} :=$ TIME() $- t_{start};$
  $\delta_{remain} := \delta_L - \delta_{transmit};$
  $\delta_w := \delta_{transmit} + \delta_{remain};$
  if $\delta_{remain} > 0$ **then**
    WAIT_MS(δ_{remain})

W	Number of windows
δ_L	Target window duration
w	Window index
λ_w	Target arrival rate
n_w	Target arrival load
t_{start}	Window start time
$\delta_{transmit}$	Transmit duration
δ_{remain}	Reamining duration
δ_w	Window duration

Algorithm 1 achieves a target arrival rate λ by transmitting tuples and waiting until the next transmission. A benchmark run is comprised of W transmission windows where λ_w can change every window. In a transmission window, n_w tuples are generated and transmitted into our SPS over $\delta_{transmit}$ ms.

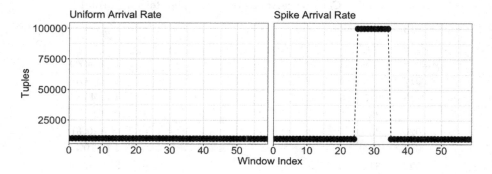

Fig. 4. Example arrival rates for uniform and spike processes

For example, if a practitioner wanted to generate a 60-second process with a uniform arrival rate of 10,000 tuples per second, the generator would set the target window arrival rate to 10000 for all windows ($\lambda_w = 10000 | \forall w \in [0, 59], L = 1000$). If, instead, that practitioner wanted to generate a 60-second process with an arrival rate of 10,000 tuples per second with a spike to 100,000 tuples per second for the middle 10 s, the generator would set the target arrival rate accordingly (($\lambda_w = 10000 | \forall w \in [0, 24] \wedge [35, 59]), L = 1000) \wedge (\lambda_w = 100000 | \forall w \in [25, 34]), L = 1000)$). Figure 4 displays the arrival rates for both cases. Throughout the remainder of this paper, we focus on uniform and spike workloads as the most parsimonious workloads capable of inducing sought-after behaviours in our system under test. Our approach is readily extensible to broader categories of the arrival process, such as step, sine and envelope-guided workloads [14].

Through our review of YCSB and NEXMark implementations, we have found that arrival rates change by modifying the arrival load n rather than the arrival duration δ. Often, δ is implicitly assumed to be 1 s. An implication of the assumption that a transmission window is analogous to 1 s is that a deviation from the target would violate the arrival rate the generator is simulating. As the arrival load is still generated and only the arrival duration is exceeded, a simple check for the total generated load at the end of an experiment would not detect this failure of achieved and target arrival rates.

Transmission windows can underrun, perfectly meet, or overrun the target window duration. In the case of underruns, Algorithm 1 waits for the remainder of the window ($\delta_{remain} = \delta_L - \delta_{transmit}$), ensuring that no windows are shorter than the target duration ($\delta_w \approx \delta_L$) (see Fig. 5). Without this, the arrival load n_w for all windows would be transmitted one after another, resulting in higher-than-expected arrival rates ($\frac{n_w}{\delta_w} > \lambda_w$) in a benchmark run and a shorter-than-expected benchmark run ($\sum_{w=0}^{W-1} \delta_w < W * \delta_L$). While Algorithm 1 has logic

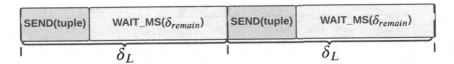

Fig. 5. Algorithm 1 maintains δ_L sized transmissions

to prevent a window underrun from occurring, there is nothing to prevent the window from overrunning (Fig. 6).

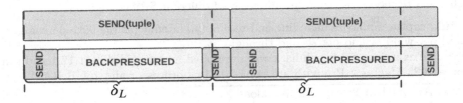

Fig. 6. High rates and backpressure induce window overruns

An overrun occurs when the time it takes to generate the load exceeds the target window duration ($\delta_{transmit} > \delta_L$). The window transmit duration $\delta_{transmit}$ is a product of the number of transmitted tuples n_w and the per-tuple transmit time. As window overruns occur when $\delta_{transmit} > \delta_L$ in §5, we will experimentally demonstrate how n_w and tuple transmit time influence $\delta_{transmit}$ and, thereby, window overruns.

5 Experimentation

Here we outline our experimental setup[3] (§ 5.1), before presenting experiments exploring the impact of high target arrival rates (§ 5.2) and pipeline backpressure (§ 5.3). We then provide concluding remarks and recommendations (§ 5.4).

5.1 Experimental Design

We implemented a closed-loop stream benchmark with three operators: a generator, a pipeline workload simulator, and a sink (see Fig. 7).

Generator: follows the basic loop from Algorithm 1, generating $n_w = \frac{\lambda_w}{L/1000}$ tuples for different target arrival rates λ_w, and waiting until the end of the transmission window.

---

[3] To support our work's broader uptake, we are developing a replication package comprising our experimental results, analysis scripts, and instructions to replicate these findings across other systems. This will be made available alongside the full paper.

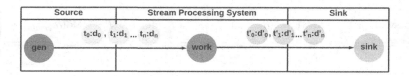

Fig. 7. Experimental benchmark simulates the system under test

Pipeline Workload Simulator: as modelled in Algorithm 2. For every ω tuples, we wait for ϵ ms to slow down the speed at which tuples pass through the pipeline. The wait frequency ω and amount of per-tuple wait time ϵ can be changed to model the compute intensity of different SPS configurations.

Sink: captures metrics including end-to-end tail latency (at the 99th percentile), and per-window and per-second throughput.

The experimental benchmark was developed in Apache Flink 1.17.1 and ran on a 2023 Macbook Pro with an Apple M2 Pro processor and 16GB RAM.

Algorithm 2: Pipeline Workload
Input ω, ϵ

$i := 0;$
while *RECIEVE(tuple)* **do**
　if $i \mod \omega = 0$ **then**
　　WAIT_MS(ϵ);
　SEND(tuple);
　$i := i + 1;$

i	Tuple index
ω	Sleep frequency
ϵ	Sleep amount

This experimental benchmark allows us to observe the effect of the generator on the benchmark run for configurable numbers of windows W, target arrival rates λ_w, and target window durations L. Similarly, we can observe the effect of the pipeline by setting the wait frequency ω and per-tuple wait time ϵ.

For the remainder of this work, set benchmark parameters are used unless otherwise specified. Generators run for 60 s with a target arrival rate of 10,000 and a target window duration of 1 sec ($Gen(W = 60, L = 1000, \lambda = 10000)$). We have selected the number of windows as it provides a 1-minute experiment and a target duration of 1 s as it's the most prevalent duration found in our code review (§3). The target rate was selected as a rate that can be met by this generator without overrunning (see Fig. 4). The pipeline simulator will be turned off unless otherwise stated ($Sim(\omega = 0, \epsilon = 0)$).

5.2 Effect of High Target Arrival Rate on Window Duration

If a window overruns, the target arrival rate is not met as the arrival load n is not transmitted within the target transmission duration ($\frac{n_w}{\delta_w} < \lambda_w$). This deviation between the target and achieved window arrival rate means that a system being benchmarked may not behave as it would in deployment under the same circumstances. To examine how window overruns occur, we must isolate changes to n_w and the time to transmit a tuple. Algorithm 1 states that the number of transmitted tuples in a window n_w is a function of the target arrival rate λ_w. By modifying λ_w we can increase n_w and observe the effects of high arrival rate on inducing window overruns.

In this experiment, we will observe the effects of high target arrival rates on window duration. If λ_w is large, the $\delta_{transmit}$ will surpass δ_L, inducing a window overrun. We can configure the benchmark to run progressively higher target arrival rates and observe the window duration δ_w to see this effect. The generator has been shown to handle arrival rates of 10,000 and 100,000 (see Fig. 4). so to induce an overrun λ_w must be set higher. By generating three processes where the middle ten windows ($w \in [25, 34]$) spike to 50x, 100x, and 150x of the baseline arrival rate of 10,000, we will see at what threshold the generator fails to meet the target arrival rate.

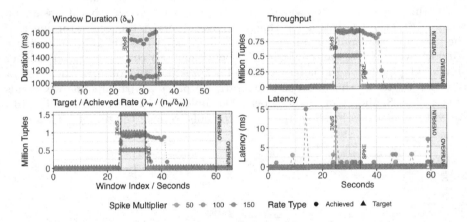

Fig. 8. High λ_w induces window overruns

In Fig. 8, the top left panel shows the window duration (δ_w), the bottom left panel shows the target and achieved window rates (λ_w, $\frac{n_w}{\delta_w}$), the top right panel shows the pipeline throughput, and the bottom right panel shows the pipeline's end-to-end per-tuple latency.

In Fig. 8, we can see that a process with a spike of 50x can be generated without inducing window overruns. The processes which spike at 100x and 150x do not reach their target arrival rates and are, on average, overrun by a factor of 1.1x and 1.7x, respectively. In the latter two processes, the overrun tuples are transmitted in subsequent windows, resulting in higher achieved arrival rates than expected. The benchmark run then overruns by the aggregate amount each window overran, with the 100x process overrunning by 1 s (100 ms * 10) and the 150x process overrunning by 7 s (700ms * 10). Throughput is almost identical to the achieved arrival rate and fails to meet the desired throughput. The pipeline latency doesn't change when the windows overrun, as the speed of generation is not affected.

By escalating the target arrival rate past what the generator can produce within the desired window duration, we have demonstrated that high target arrival rates induce window overruns. Benchmarking practitioners should ensure that the target arrival rate provided to a generator is within that generator's capacity to produce, or else the failure to achieve the desired throughput will partially result in the generator rather than the system under test.

5.3 Effect of Pipeline Backpressure on Window Duration

In §2, we state that closed-loop stream generators can be influenced by the system under test. Backpressure can propagate up the pipeline and throttle the rate at which tuples are transmitted into the pipeline, increasing the per-tuple transmit time. By increasing the pipeline workload, we can induce backpressure and observe the effects of tuple transmit time on window overruns.

In this experiment, we will observe the effect of pipeline backpressure on window duration. Given a λ_w that we have observed the generator meet, if the per-tuple transmit time increases, the $\delta_{transmit}$ will surpass δ_L, inducing a window overrun. We can configure the benchmark to run progressively higher simulated pipeline workload to trigger backpressure and observe the window duration δ_w to see this effect. By generating a process with pipeline workloads of 0, 1, 2, 4, 8, and 16 ms of sleep per 1000 tuples ($\omega = 1000, \epsilon \in 0, 1, 2, 4, 8, 16$), we will see at what threshold the generator fails to meet the target arrival rate.

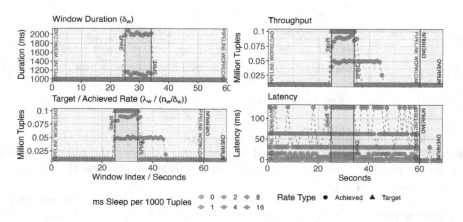

Fig. 9. Backpressure from pipeline workload induces window overruns

In Fig. 9, we can see that under a low pipeline workload ($\epsilon \in 0, 1, 2, 4$), the process can be generated without inducing window overruns. Under higher workloads $\epsilon \in 8, 16$), the processes do not reach their target arrival rates and are, on average, overrun by a factor of 1.1x and 2x, respectively. Similarly to the high target arrival rates, the higher workloads process overran by the aggregate amount of each window overrun, with the $\epsilon = 8$ process overrunning by

1 s (100ms * 10) and the $\epsilon = 16$ process overrunning by 10 s (1000ms * 10). Throughput matches the achieved arrival rate and fails to meet the desired throughput. Latency increases with increased pipeline workload, which, as the workload increases the time it takes for a tuple to pass through the pipeline, logically follows.

By escalating the pipeline workload we have demonstrated that backpressure propagates to the generator and throttles the rate at which tuples are generated. If the degradation in throughput was due to the workload itself and not the backpressured generator, then the degradation would only be seen in the throughput metric and not in the achieved arrival rate. Benchmarking practitioners should ensure that backpressure is not affecting the generator, or else the failure to achieve the desired throughput will partially be the result of backpressure rather than the system under test.

5.4 Key Takeaways

High target arrival rates and backpressure from pipeline workload can increase $\delta_{transmit}$ and cause a window to overrun. Overruns cause the achieved window arrival rate to fall short of the target and underload the system, leading to the system being tested to perform differently to deployment. Additionally, tuples that the generator is not able to generate in an overran window spill over into subsequent windows, resulting in further arrival rate mismatches. Without understanding the interplay of high arrival rates or backpressure, the effects of window overruns can pass unnoticed.

In our experiments, we isolate these effects, but in a benchmark, they can be explained by the performance of the system under test. This explainability can be an issue - as it may not be clear that benchmarkers should check if the benchmark itself is operating correctly. To avoid window overruns from influencing benchmark results, practitioners should determine the maximum achievable arrival rates and monitor their systems for backpressure. Alternatively, simply monitoring window duration provides a litmus check that the generator operated correctly in a benchmark run.

Our findings highlight the need for practitioners to proactively measure the effects of backpressure during benchmark runs, as is often done in established benchmark suites [32]. A simple validation checklist for the loop in Algorithm 1 could be: **a)** Ensure transmission windows finish within 1% of target window duration, **b)** Ensure achieved arrival rates are within 5% of target arrival rates.

6 Mitigating Overruns

We have demonstrated in Sect. 5 that current approaches to benchmarking are susceptible to overruns due to high target arrival rates, and due to pipeline backpressure. This leads to a mismatch between the achieved and target window durations ($\delta_w \neq \delta_L$). Consequently, the effective generation rate observed by the pipeline is lessened, and the total runtime for a benchmark increases.

We now seek to demonstrate the impact of terminating each window at the end of the window duration target δ_L. This mechanism is analogous to load shedding, which is experienced in the open-loop deployment of an SPS system. In Fig. 10 we implement the experiments from § 5.2 and § 5.3 but cutoff windows when they exceed δ_L. Figure 11 demonstrates that this mechanism is effective at almost completely mitigating window overruns across a benchmark run. We can see that the arrival rate mismatch does not propagate into the subsequent windows, as in Figs. 9 and 8. Furthermore, the assumption from §4 that a transmission window is analogous to a second is now true, as the experiment now runs for a total of 60 s without exceeding.

We believe our cutoff mechanism is a promising direction for benchmarking, alongside proactive monitoring of transmit duration $\delta_{transmit}$ across an experiment. Specifically, this mechanism allows us to not only measure the number of windows that violate the target length but also establish thresholds of acceptable behaviour based on our use case. Similar approaches are adopted by benchmark consortia, such as SPEC, in their Run and Reporting rules [32], e.g. *"Verify that the elapsed time for each measurement interval is at least 99.5% but no more than 101% of the configured interval length."*.

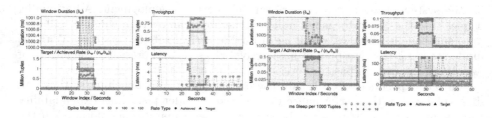

Fig. 10. Cutoff prevents overrun spillage and ends the benchmark on time

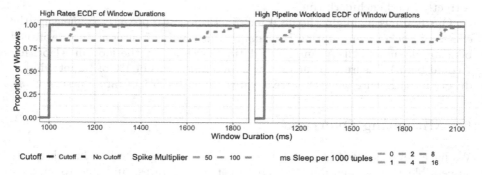

Fig. 11. Cutoff effectively prevents window overruns for all windows

7 Threats to Validity

Here, we frame the scope and limitations of this work with respect to *construct*, *internal* and *external* validity [38].

Construct Validity: We evaluate system performance in relation to throughput and latency. The performance of workload generator as the proportion of target workload generated and window duration. Future work will explicitly model the window cutoff's impact on the trajectory and extent of backpressure.

Internal Validity: Ongoing work seeks to clearly distinguish the halting as a consequence of the backpressure mechanism, from halting behaviours caused by other external factors (e.g. IO contention or interrupts). Furthermore, our system model does not currently capture that the computational cost of stateful operators scales non-linearly with state size [24,25].

External Validity: While our current experiments centre around a single workload for Apache Flink, our approach can be applied broadly across a range of systems. The experimental work here relies only on measures which are provided as standard by many contemporary streaming platforms [14,15].

8 Conclusion

We have investigated benchmarking practices that underpin a large proportion of performance studies on streaming systems. We have explored the implications of benchmark designs, particularly with respect to their robustness in the presence of self-regulating mechanisms such as backpressure. We present empirical findings which demonstrate that high generation rates and backpressure may lead to window overruns in benchmarks, significantly threatening the validity of results. We make several suggested amendments to benchmarking practices to mitigate their impact, including the collection of more fine-grained metrics on system performance, and the need for proactive validation checks.

We propose further work in the development of generator spacing algorithms, analysing the impact of the frequency of checking for cutoff on benchmark runs, and generalised metrics to measure backpressure. While the transmit all messages and wait behaviour in Algorithm 1 does produce a process with appropriately spaced messages to meet the target arrival rate, the transmit utilisation is limited to the beginning of the window. This behaviour does not accurately reflect how tuples arrive at an SPS within a second, so developing methods that space out transmit throughout a window provides a more realistic benchmark. Checking for overruns at every tuple introduces additional overhead to the generator, and through experimentation we can determine the optimum cutoff frequency to minimise window overruns whilst minimising overhead. Finally, backpressure is a mechanism that we have demonstrated has a significant effect on closed-loop benchmarks, but metrics to monitor backpressure are specific to each SPS. A generalised metric to measure the presence of backpressure would allow practitioners to quantify its impact on benchmark behaviour and respond accordingly.

References

1. Abedi, A. and Brecht, T.: Conducting repeatable experiments in highly variable cloud computing environments. In: ACM/SPEC ICPE, pp. 287–292 (2017)
2. Alves, L., Veiga, L.: Stream economics: resource efficiency in streams with task over-allocation and load shedding. In: Martins, R., Selimi, M. (eds.) Distributed Applications and Interoperable Systems. DAIS 2024, pp. 1–17. Springer-Verlag, Cham (2024). https://doi.org/10.1007/978-3-031-62638-8_1, ISBN 978-3-031-62637-1
3. Bartolomeo, G., Cao, J., Su, X., Mohan, N.: Characterizing distributed mobile augmented reality applications at the edge. pp. 9 – 18 (2023). Cited by: 0
4. Bouckaert, S., Gerwen, V.V., Moerman, I., Phillips, S., Wilander, J.: BONFIRE: benchmarking computers and computer networks (2011)
5. Chantzialexiou, G., Luckow, A., Jha, S., Pilot-streaming: a stream processing framework for high-performance computing, pp. 177–188 (2018). Cited by: 12. All Open Access, Green Open Access (2018)
6. Cooper, B.F., Silberstein, A., Tam, E., Ramakrishnan, R., Sears, R.: Benchmarking cloud serving systems with YCSB. In: ACM SoCC 2010, pp. 143–154 (2010). ISBN 9781450300360
7. Duplyakin, D., et al.: Avoiding the ordering trap in systems performance measurement. In: 2023 USENIX Annual Technical Conference, pp. 373–386 (2023)
8. Eismann, S., et al.: A case study on the stability of performance tests for serverless applications. J. Syst. Softw. **189**, 111294 (2022)
9. Fu, X., Ghaffar, T., Davis, J.C., Lee, D.: Edgewise: a better stream processing engine for the edge, pp. 929 – 945 (2019). Cited by: 54
10. Gautam, B., Basava, A.: Performance prediction of data streams on high-performance architecture. Human-centric Comput. Inf. Sci. **9**(1), 2 (2019). Cited by: 9. All Open Access, Hybrid Gold Open Access (2019)
11. Huang, X., Shao, Z., Yang, Y.: POTUS: predictive online tuple scheduling for data stream processing systems. IEEE Trans. Cloud Comput. **10**(4), 2863–2875 (2022). https://doi.org/10.1109/TCC.2020.3032577. Cited by: 4; All Open Access, Green Open Access
12. Jamieson, S.: Dynamic scaling of distributed data-flows under uncertainty. In: ACM DEBS, pp. 230–233 (2020)
13. Jamieson, S., Forshaw, M.: Measuring streaming system robustness using non-parametric goodness-of-fit tests. In: Gilly, K., Thomas, N. (eds.) Computer Performance Engineering. EPEW 2022, pp. 3–18. Springer, Cham (2022). https://doi.org/10.1007/978-3-031-25049-1_1
14. Jamieson, S. and Forshaw, M.: On improving streaming system autoscaler behaviour using windowing and weighting methods. In: ACM DEBS, pp. 68–79 (2023)
15. Kalavri, V., Liagouris, J., Hoffmann, M., Dimitrova, D., Forshaw, M., Roscoe, T.: Three steps is all you need: fast, accurate, automatic scaling decisions for distributed streaming dataflows. In: USENIX OSDI 2018, pp. 783–798 (2018)
16. Kallas, K., Niksic, F., Stanford, C., Alur, R.: DiffStream: differential output testing for stream processing programs. In: Proceedings of the ACM on Programming Languages 4(OOPSLA) (2020). https://doi.org/10.1145/3428221
17. Karimov, J., Rabl, T., Katsifodimos, A., Samarev, R., Heiskanen, H., Markl, V.: Benchmarking distributed stream data processing systems. In: 2018 IEEE 34th International Conference on Data Engineering (ICDE), pp. 1507–1518. IEEE (2018)

18. Kogias, M., Mallon, S., Bugnion, E.: Lancet: a self-correcting latency measuring tool. In: 2019 USENIX Annual Technical Conference, pp. 881–896 (2019). ISBN 978-1-939133-03-8
19. Li, B., Zhang, Z., Zheng, T., Zhong, Q., Huang, Q., Cheng, X.: Marabunta: continuous distributed processing of skewed streams. In 2020 20th IEEE/ACM International Symposium on Cluster, Cloud and Internet Computing (CCGRID), pp. 252–261. IEEE. Marabunta: Continuous distributed processing of skewed streams. In: IEEE/ACM International Symposium on Cluster, Cloud and Internet Computing, pp. 252 – 261 (2020). Cited by: 4
20. Lian, J., et al.: ContTune: continuous tuning by conservative Bayesian optimization for distributed stream data processing systems. Proc. VLDB Endowment **16**(13), 4282–4295 (2023). Cited by: 1. All Open Access, Green Open Access (2023)
21. Lu, P., Yue, Y., Yuan, L., Zhang, Y.: AutoFlow: hotspot-aware, dynamic load balancing for distributed stream processing. LNCS 13157, pp. 133–151 (2022). https://doi.org/10.1007/978-3-030-95391-1_9
22. Maricq, A., Duplyakin, D., Jimenez, I., Maltzahn, C., Stutsman, R., Ricci, R.: Taming performance variability. In: USENIX OSDI, pp. 409–425 (2018)
23. Ntoulias, E., Alevizos, E., Artikis, A., Koumparos, A.: Online trajectory analysis with scalable event recognition, vol. 2841 (2021). Cited by: 2
24. Omoregbee, P., Forshaw, M., Thomas, N.: A state-size inclusive approach to optimizing stream processing applications. In: EPEW, pp. 325–339 (2023)
25. Omoregbee, P., Thomas, N., Forshaw, M.: Analyzing performance effects of window size on streaming operator throughput. In: UKPEW, p. 18 (2023)
26. Papadopoulos, A.V., et al.: Methodological principles for reproducible performance evaluation in cloud computing. IEEE Trans. Softw. Eng. **47**(8), 1528–1543 (2019)
27. Prisyazhnyy, I.: On coordinated omission (2021). https://www.scylladb.com/2021/04/22/on-coordinated-omission/
28. Prokopec, A.: Encoding the building blocks of communication, pp. 104–118 (2017). https://doi.org/10.1145/3133850.3133865
29. Qu, W., Dessloch, S.: A lightweight elastic queue middleware for distributed streaming pipeline. In: Bellatreche, L., Chakravarthy, S. (eds.) DaWaK 2017. LNCS, vol. 10440, pp. 173–182. Springer, Cham (2017). https://doi.org/10.1007/978-3-319-64283-3_13
30. Schroeder, B., Wierman, A., Harchol-Balter, M.: Open Versus Closed: A Cautionary Tale. vol. 3. USENIX NSDI (2006)
31. SIGPLAN. SIGPLAN empirical evaluation checklist. https://www.sigplan.org/Resources/EmpiricalEvaluation/
32. Standard performance evaluation corporation. Specpower_ssj2008 run and reporting rules. https://www.spec.org/power/docs/SPECpower_ssj2008-Run_Reporting_Rules.html#2.1
33. Tassiulas, L., Ephremides, A.: Stability properties of constrained queueing systems and scheduling policies for maximum throughput in multihop radio networks. In: 29th IEEE Conference on Decision and Control, pp. 2130–2132. IEEE (1990)
34. The Turing Way: Definitions. https://the-turing-way.netlify.app/reproducible-research/overview/overview-definitions.html
35. Tucker, P., Tufte, K., Papadimos, V., Maier, D.: Nexmark - a benchmark for queries over data streams draft (2002)
36. Wakart, N.: Correcting YCSB's coordinated omission problem (2015). https://psy-lob-saw.blogspot.com/2015/03/fixing-ycsb-coordinated-omission.html
37. Winter, S., et al.:A retrospective study of one decade of artifact evaluations. In: ESEC, FSE, pp. page 145–156 (2022). ISBN **9781450394130**,(2022)

38. Wohlin, C., Runeson, P., Höst, M., Ohlsson, M.C., Regnell, B., Wesslén, A.: Experimentation in Software Engineering. Springer Science & Business Media (2012). https://doi.org/10.1007/978-3-662-69306-3
39. Ye, Q., Liu, W., Wu, C.Q.: NoStop: a novel configuration optimization scheme for spark streaming (2021). https://doi.org/10.1145/3472456.3472515
40. Zilberman, N., Moore, A.W.: Thoughts about artifact badging. SIGCOMM Comput. Commun. Rev. **50**(2), 60–63 (2020). ISSN 0146-4833

Author Index

B
Barbierato, Enrico 60
Bertocci, Nicola 14

C
Capra, Lorentzo 106
Carnevali, Laura 14

D
Dixon, Iain 137

E
Ezhilchelvan, Paul 1, 45

F
Forshaw, Matthew 137

G
Gatti, Alice 60
Gribaudo, M. 121
Gribaudo, Marco 60, 106

H
Horváth, András 75

I
Iacono, Mauro 60

L
Laghbi, Hassan 29
Liu, Ye 1

M
Malakhov, Ivan 91
Marin, Andrea 91
Matthews, Joe 137
Mitrani, Isi 1

P
Paolieri, Marco 75
Phung-Duc, T. 121
Piazza, Carla 91

R
Rossi, Sabina 91

S
Scommegna, Leonardo 14
Smuseva, Daria 91

T
Thomas, Nigel 29

V
Vicario, Enrico 14, 75

W
Wang, Yingming 45
Waudby, Jack 45
Webber, Jim 45